CONTEMPORARY DIGITAL TRANSFORMATION AND ORGANIZATIONAL EFFECTIVENESS IN BUSINESS 4.0

CONTEMPORARY DIGITAL TRANSFORMATION AND ORGANIZATIONAL EFFECTIVENESS IN BUSINESS 4.0

Nilanjan Ray, PhD
Sudin Bag, PhD
Tapas Kumar Chatterjee, PhD
Editors

AAP | APPLE ACADEMIC PRESS

First edition published 2025

Apple Academic Press Inc.
1265 Goldenrod Circle, NE,
Palm Bay, FL 32905 USA

760 Laurentian Drive, Unit 19,
Burlington, ON L7N 0A4, CANADA

CRC Press
2385 NW Executive Center Drive,
Suite 320, Boca Raton FL 33431

4 Park Square, Milton Park,
Abingdon, Oxon, OX14 4RN UK

Apple Academic Press exclusively co-publishes with CRC Press, an imprint of Taylor & Francis Group, LLC

Library and Archives Canada Cataloguing in Publication

CIP data on file with Canada Library and Archives

Library of Congress Cataloging-in-Publication Data

CIP data on file with US Library of Congress

ISBN: 978-1-77491-654-4 (hbk)
ISBN: 978-1-77491-655-1 (pbk)
ISBN: 978-1-00350-501-3 (ebk)

About the Editors

Nilanjan Ray, PhD
Associate Professor, Department of Management Studies, JIS University, West Bengal, India

Nilanjan Ray, PhD, is from Kolkata, India, and presently associated as an Associate Professor in the Department of Management Studies at JIS University, West Bengal, India. Prior to joining JIS University, Dr. Ray was at the Institute of Leadership Entrepreneurship and Development as Associate Professor and Head of Department with additional responsibility as Director IQAC. Prior to that he was also associated at Adamas University as Associate Professor of Marketing Management and Centre Coordinator for Research in Business Analytics at Adamas University in the Department of Management, School of Business and Economics, West Bengal, India. Dr. Ray has obtained a certified an Accredited Management Teacher Award from the All India Management Association, New Delhi, India. He has earned the degrees of PhD (Mktg), MCom (Mktg), MBA (Mktg), and STC FMRM (IIT-KGP). He has more than 12 years of teaching and six years of research experience. He has supervised and awarded two doctoral scholars and guided around 56 postgraduate students projects also. Dr. Ray has contributed over 90 research papers in reputed national and international referred, peer-reviewed journals and proceedings as well as 13 edited research handbooks from Springer, IGI-Global USA, and Apple Academic Press/CRC Press (A Taylor & Francis Group), USA. He has obtained one patent from Germany and has two copyrights from India. He has also associated himself as a reviewer for Tourism Management, Journal of Service Marketing, Journal of Business and Economics, and Research Journal of Business and Management Accounting and as an editorial board member of several refereed journals. Dr. Ray has organized several faculty development programs, national and international conferences, and management doctoral colloquiums. Dr. Ray is a life member of the International Business Studies Academia as well as a fellow member of the Institute of Research Engineers and Doctors, the Universal Association of Arts And Management Professionals (UAAMP) New York, USA, and Calcutta Management Association (CMA).

Sudin Bag, PhD
Assistant Professor (Stage-2) and Ex-Head in the Department of Business Administration, Vidyasagar University Midnapore, West Bengal, India

Sudin Bag, PhD, is currently working as an Assistant Professor (Stage-2) and is Ex-Head of the Department of Business Administration, Vidyasagar University Midnapore, West Bengal, India. Dr. Bag has more than 11 years of working experience which is inclusive of corporate exposure as well as teaching management at colleges and university. His areas of interest include marketing management, services marketing, research methodology, consumer behavior, and international marketing. Dr. Bag has published over 28 research papers in reputed national and international journals and has contributed a good number of research papers in edited volumes including with Springer Nature, Emerald, etc. He has also authored two books on management. Dr. Bag has attended and presented over 42 research papers at national and international conferences in India and visited several countries for academic activities. He has acted as a resource person in various academic assignments, such as for seminars, conferences, and short-term courses. Dr. Bag is a life member of the International Business Studies Academia, Durgapur, and All India Accounting Association, Kolkata branch, India. Dr. Bag received the IARA Best Faculty Award (2019) in the field of Management on April 21, 2019, conferred by the Indian Academic Researchers Association in Tiruchirappalli, Tamil Nadu. He has also been awarded and completed a research project from ICSSR under the special call for studies focusing on the social science dimensions of the COVID-19 coronavirus pandemic.

Tapas Kumar Chatterjee, PhD
Associate Professor, Marketing Area, IMT Nagpur, India

Tapas Kumar Chatterjee, PhD, is currently working as an Associate Professor in the Marketing Area at IMT Nagpur. He holds an MBA in Marketing from Patna University and a PhD in Management from National Law University, Jodhpur. With 20 years of corporate experience and 15 years in academics, he has published numerous research articles in national and international journals. He has also presented several papers at national and international conferences. Additionally, he is the author of a book and actively engages in networking. Dr. Chatterjee has delivered talks as a plenary speaker, guest of honor, keynote speaker, and more. In his leisure time, he enjoys singing karaoke and playing chess.

Contents

Contributors

Nadia Adnan
Assistant Professor Management and Marketing College of Business Administration, Prince Mohammad Bin Fahd University, Dhahran, Saudi Arabia

Shivaji Banerjee
Senior Assistant Professor in Management, PG and Research Department of Commerce, St. Xavier's College Autonomous; and Visiting Faculty, University of Calcutta and IISWBM, Kolkata, West Bengal, India

Debasish Batabyal
Assistant Professor, Amity University Kolkata, West Bengal, India

Archie Behuria
Post-Graduate and Research Department of Commerce, St. Xavier's College Autonomous, Kolkata, West Bengal, India

Indranil Bose
Vice President-Academics and Professor, School of Business and Economics, Adamas University, Kolkata, West Bengal, India

Mahul Brahma
Professor and Dean, School of Media and Communication (SoMC), Adamas University, Barasat, Kolkata, West Bengal, India

Namrata Chakrabortty
Assistant Professor, Guru Nanak Institute of Technology, Kolkata, West Bengal, India

Mainak Chakraborty
Associate Professor, Department of Business Management, School of Management, Swami Vivekananda University, West Bengal, India

Pratim Chatterjee
Assistant Professor, Amity University Kolkata, West Bengal, India

Mrinal Kanti Das
Assistant Professor, Department of Commerce, Kanchrapara College, Kanchrapara, West Bengal, India

Sulagna Das
Associate Professor, JIS University, Kolkata, West Bengal, India

Shelly De (Pandit)
Associate Professor, Department of Commerce, Bijoy Krishna Girls' College, Howrah, West Bengal, India

Abhijit Dutta
Professor, Sikkim University, Gangtok, Sikkim, India

Hafsa Fahad
UNSW Student, University of New South Wales, Australia

Anjan Kumar Ghosh
Research Scholar, Department of Business Administration (Human Resource), The University of Burdwan, West Bengal, India

Arindam Ghosh
Associate Professor, Department of Management Science and Humanities, MCKV Institute of Engineering, West Bengal, India

Seyed Farhad Hosseini
Department of Business Management and Accounting, Faculty of Management and Accounting, Razi University, Kermanshah, Iran

Amir Jafar
Associate Professor, Department of Business Administration, Aligarh Muslim University, Murshidabad Center, West Bengal, India

Rachana Jaiswal
Assistant Professor, Department of Business Management, HNB Garhwal (A Central) University, Srinagar, Uttarakhand, India

Areej Jamil
Salim Habib University, Karachi, Pakistan

Siddhartha Karak
Post-Graduate and Research Department of Commerce, St. Xavier's College Autonomous, Kolkata, West Bengal, India

Sharmila Kayal
Associate Professor, Department of Journalism and Communication, School of Media and Communication, Adamas University, West Bengal, India

Dhanonjoy Kumar
Professor, Department of Management, Islamic University, Kushtia, Bangladesh

Amrita Majumdar
Associate Professor, Amity Business School, Amity University, Kolkata, West Bengal, India

Sudipta Majumdar
Associate Professor, School of Business and Economics, Adamas University, Kolkata, West Bengal, India

Sumedha Majumder
Assistant Professor, Department of Commerce, Shree Agrasain College, Howrah, West Bengal, India

Fakhraddin Maroofi
Department of Management, Faculty of Humanities and Social Sciences, University of Kurdistan, Sanandaj, Iran

Soumya Mukherjee
Associate Professor, Department of Management Studies, Techno India (Hooghly Campus), Chinsura, Hooghly, West Bengal, India

Swati Pal
Associate Professor, Department of Management, School of Business, Adamas University, Barasat, Kolkata, West Bengal, India

Priyajit Ray
Research Scholar, Department of Commerce, University of Calcutta, Kolkata, West Bengal, India

Kumail Raza
Salim Habib University, Karachi, Pakistan

Omar C. Vargas-González
Departamento de Sistemas, Instituto Tecnológico Nacional de México, Cd, Guzmán

José Vargas-Hernández
Research Professor, Institute of Technological Postgraduate and Research Division, Instituto Tecnológico Mario Molina, Zapopan, Mexico

Foreword

This is a collection of chapters skillfully assembled by the three editors, covering a wide range of topics that are of interest to modern researchers and managers. Each of the 13 chapters presents cutting-edge research findings on a subject that will captivate all readers.

Undoubtedly, sustainability stands out as the most crucial aspect of business transformation that we are currently witnessing. Entrepreneurs face the ongoing challenge of establishing sustainable new businesses. Existing businesses, on the other hand, must strive to remain relevant while adhering to the new societal norms. In today's world, pursuing growth at any cost is no longer acceptable to governments and society.

These are just a few of the issues that the various authors address in their papers, presenting original research in these areas or related fields.

Technology also plays a significant role, particularly with the widespread use of Artificial Intelligence across various domains. Its applications are vast and continuously evolving. Cryptocurrencies are built on blockchain principles, while Chat GPT poses a threat to search engines like Google. Virtually, every innovation is rooted in technology, with very few exceptions.

Understanding these issues is crucial for business transformation, and incorporating strategies to leverage technology is essential for all public and private organizations.

Therefore, I wholeheartedly welcome this collection of timely and important papers in the form of diverse book chapters, carefully selected and curated by the editors. They will undoubtedly prove immensely valuable to both corporate and academic readers.

—Rajendra Nargundkar
Vice Chancellor, Prestige University

Preface

Sustainability has become a buzzword in the business world today, characterized by transformation. Building sustainable new businesses is an omnipresent challenge faced by entrepreneurs. For existing businesses, remaining both relevant and conforming to new norms expected from society poses a real dilemma. Growth at all costs is no longer acceptable to the government and society in the current era.

These are some of the issues that various authors deal with in these chapters, presented through their original research into one of these, or related, areas.

This volume covers various interdisciplinary areas such as sustainable development goals (SDG) for organizational effectiveness in the VUCA (volatility, uncertainty, complexity, and ambiguity) world, interrelationships and implications of socio-intercultural sustainable natural resources management and governance, entrepreneurial marketing practices, Bitcoin, cryptocurrencies, digital education and students' attitudes, current applications of AI machine learning in mainstream practices, digital financial services, the agri-food industry of Pakistan, specialized technical education and entrepreneurism, artificial intelligence in the road map of careers in the financial sector, digital ecosystems in a VUCA world, and the role of information systems.

This book, *Contemporary Digital Transformation and Organizational Effectiveness in Business 4.0*, is a collection of timely and important chapters that deal with these topics. The editors have carefully selected and curated these chapters, which will be immensely useful to readers, both from corporate and academic backgrounds.

CHAPTER 1

Accomplishing Sustainable Development Goals for Organizational Effectiveness in a VUCA World

SIDDHARTHA KARAK,[1] ARCHIE BEHURIA,[1] and SHIVAJI BANERJEE[2]

[1]*Post-Graduate and Research Department of Commerce, St. Xavier's College Autonomous, Kolkata, West Bengal, India*

[2]*Senior Assistant Professor in Management, PG and Research Department of Commerce, St. Xavier's College Autonomous; and Visiting Faculty, University of Calcutta and IISWBM, Kolkata, West Bengal, India*

ABSTRACT

Volatility, uncertainty, complexity, and ambiguity (VUCA) can be seen as a way to enhance leadership and management skills. Threats can be transformed into opportunities if organizations begin to prioritize areas such as decision-making, innovation, implementation, creativity, organizational culture, and dynamics. Being slow and outdated is not an option in this VUCA world. It requires newer, practical, and innovative approaches to thrive in a VUCA environment.

The objective of this study is to explore the different factors that influence a VUCA world. It aims to analyze the connections between these factors and the 17 sustainable development goals (SDGs) established by the United Nations (UN) in 2016, in order to create a sustainable environment and address the challenges of the VUCA world through various solution models.

Contemporary Digital Transformation and Organizational Effectiveness in Business 4.0.
Nilanjan Ray, Sudin Bag, & Tapas Kumar Chatterjee (Eds.)

1.1 INTRODUCTION

VUCA stands for volatility, uncertainty, complexity, and ambiguity. It represents a set of four distinct challenges that are commonly encountered in modern society. Volatility presents difficulties in terms of the speed and unpredictable nature of change. Uncertainty refers to a lack of understanding of variables. Complexity arises from the multiplicity and interconnectedness of factors, leading to chaos and confusion. Lastly, ambiguity refers to the haziness caused by a lack of clarity about events.

Over the years, VUCA has had a profound impact on society as a whole, and particularly on the business world. The business environment has shifted from one of certainty, stability, and familiarity to one of constant fluctuation. VUCA is, in a sense, unavoidable. Therefore, it necessitates a strategic framework known as 'VUCA Prime' to effectively respond to threats in the VUCA world. VUCA Prime proposes four key responses:

- Volatility with vision;
- Uncertainty with understanding;
- Complexity with clarity; and
- Ambiguity with agility.

Similarly, the observe, orient, decide, and act (OODA) loop is a popular four-step model to counter threats and sustain in a VUCA world. Nowadays, with advancements in science and technology, innovations occur more frequently than ever, almost following Moore's Law. According to Moore's Law, the number of innovations doubles approximately every two years. Gordon Moore first observed this law in 1965, stating that "The number of transistors in a dense integrated circuit doubles about every two years" (Gianfagna, 2021).

The purpose of this study is to analyze the game changers of a VUCA world and their role in accomplishing the 17 sustainable development goals (SDGs) framed by the United Nations (UN) in 2016 to build a sustainable environment. Additionally, a solution model called the Identify, Prioritize, Analyze, and Deduce (IPAD) model is being developed to counter the challenges of a VUCA world and assist these game-changers in creating a sustainable environment.

1.2 LITERATURE REVIEW

After evaluating and closely analyzing the existing literature on the factors that affect a VUCA environment, extracts from the most relevant sources were used in the secondary data analysis of this chapter.

1.2.1 EVIDENCE-BASED POLICY MAKING (LORRAINE EDEN, 2021)

One of the notable sources is the paper titled "Evidence-Based Policy Making in a VUCA World" by Lorraine Eden, Charles F. Hermann, and Stewart R. Miller, published in 2021. In their research, they concluded that to assist policymakers in using evidence-based policies in a VUCA world, researchers should provide insights drawn from event analysis. They emphasized the importance of understanding the nature of global shocks and situating them within event analysis. Additionally, they highlighted the need to choose appropriate event-analysis research methods for analyzing the antecedents and consequences of these shocks.

1.2.2 CONNECTIVITY AND SDG9 IN THE AGE OF VUCA (ANAND, 2020)

In November 2020, Jaimal Anand presented a paper at the Pathfinder Indian Ocean Security Conference titled "Connectivity in an Age of Volatility, Uncertainty, Complexity, and Ambiguity (VUCA): Realizing SDG9 amid the COVID pandemic." The paper was presented in the context of the looming 'global reset' as developing nations continued to bear the economic brunt of underdevelopment that was brutally exposed due to the pandemic. In his conclusion, Anand emphasized the role of innovation and infrastructure in promoting new technologies, facilitating international trade, and enabling the efficient use of resources. The chapter also discusses 9 goal-oriented targets that can help sustain and develop the industry in the upcoming days.

1.2.3 HIGHER EDUCATION IN A VUCA WORLD (KORSAKOVA, 2020)

In 2020, Tatiana V. Korsakova wrote a paper titled "Higher Education in VUCA-World: New Metaphor of University." The paper elaborates on specific features of the new reality that are significant for building modern organizational systems in universities. It concludes that the acronym VUCA, which stands for Validation, Uniqueness, Complementarity, and Attractiveness, represents the key aspects of this new reality.

1.2.4 DEVELOPING COMPETENCIES IN VUCA WORLD (ELENA SHKARUPETA, 2020)

In 2020, Elena Shkarupeta and a team of researchers from Russia wrote a research paper titled "Strategies for Developing Competencies for the VUCA

World." In this paper, they explored techniques, tools, and procedures for establishing important circumstances for training staff in the digital region economy and improving the educational system. The goal was to generate a competitive workforce for the digital region economy. The COVID-19 pandemic in 2020 brought extreme alterations that all educational systems worldwide had to deal with, resulting in shock-born innovations. The need for change can have both positive and negative repercussions for educational institutions, but it can also be a source of creativity. Significant advances in education in 2020 included adjustment technologies, coping tactics, and mutual support mechanisms in the networked community. Thanks to mutual micro-learning processes, unparalleled mutual attention among stakeholders, and effective resource distribution, an immediate transition to an online environment became conceivable.

1.2.5 COORDINATED GOVERNANCE AND VUCA (MORENO, 2020)

In 2020, Oscar Mauricio Covarrubias Moreno authored a paper titled "Coordinated Governance in the VUCA Scenario," which addressed the importance of coordination in the study and practice of governance. The paper concluded that coordination and governance are interconnected concepts that mutually inform and enhance each other. Consequently, establishing a robust coordination structure is crucial for effectively implementing policies and programs aimed at addressing social issues. In a world where public problems are increasingly complex and interconnected, coordination also plays a vital role in promoting coherent actions, thereby presenting a significant governance challenge.

1.2.6 DARK SIDE OF VUCA (NATHAN BENNETT, 2014)

In 2014, Nathan Bennett, James Lemoine, and three co-authors presented a paper titled "What a Difference a Word Makes: Understanding Threats to Performance in a VUCA World." In their study, they concluded that globalization has brought both opportunities and threats. The ongoing global recession has dampened hopes for a return to prosperity, while the challenges of adapting to technological changes and navigating worker demographic transitions have further complicated matters for executives. Recognizing and addressing the multitude of issues that can undermine organizational effectiveness is increasingly challenging.

1.2.7 LEADERSHIP MANDATES IN A VUCA WORLD (SURYANARAYANA, 2022)

In 2022, Prof. A. Suryanarayana authored a paper titled "Insights into Leadership Mandates and Imperatives in a VUCA World." In this paper, he concluded that a VUCA society poses threats to effectiveness, efficiency, and profitability, including volatile environments, increased competition, and evolving client needs. Maintaining consistency across organizational dimensions such as technology, strategy, culture, and processes remains an ongoing challenge. In today's VUCA world, ensuring the health and sustainability of firms is a complex task. Leaders in a VUCA society must adopt a systematic approach to implementing behavioral science concepts and practices to enhance individual and organizational effectiveness. The focus of leadership should be on organizational efficacy and improving how they operate through comprehensive system reform. Their primary emphasis is on taking action and achieving desired outcomes through pre-planned activities.

1.3 CONCEPTUAL FRAMEWORK

Sustainable development can be defined as the concept in which mankind should sustain resources to meet current needs, while also ensuring that future resource needs are not compromised. In the past three decades, with the emergence of the VUCA world, there has been significant growth in business and the economy on a large scale. To address these undesirable changes, the UN has designed 17 goals known as the SDGs. The purpose of these SDGs is to achieve the targets by 2030, with the aim of creating a greener, healthier, and more peaceful world to live in.

There are essentially four factors that challenge and at the same time, act as game-changers in a VUCA world:

1. **International Alliances and Connectivity:** To address the threats of the VUCA world, societies must collaborate. International alliances play a significant role in integrating the societies and economies of the world through various agreements, memoranda of understanding (MOUs), and the formation of alliances. With the increase in international conflicts and discourses over the years, international alliances have a crucial role to play in ensuring sustainability through policy framing and implementation.

2. **Entrepreneurship:** The role of entrepreneurs in a VUCA world is pivotal. Entrepreneurs are one of the major driving forces of an economy. They not only provide employment opportunities but also contribute to economic development, connectivity, and sustainability.
3. **Education:** The educational sector is the most fundamental factor in an economy. As the world progresses through various phases of the industrial revolution and faces the challenges of a VUCA world, the educational system undergoes changes and developments. These developments form the foundation upon which a whole new generation is built. With advancements in science, technology, and the economy, the educational sector faces intense competition among students for higher ranks and better employment opportunities. Thus, education plays a major role in building a sustainable environment.
4. **Technology:** Technological advancement has accelerated paradigm transitions, resulting in a more diversified environment and helping to shape our VUCA reality. The second, third, and fourth waves of economic development are all present at the same time. Some societies, countries, industries, firms, and individuals may still be operating in the second paradigm, while others might be operating in the third or fourth paradigm, or a combination of the three – the complexity of our modern world is that the different paradigms are running in parallel.

   Society can only ensure sustainable development when it has successfully dealt with the challenges of vulnerability, uncertainty, complexity, and ambiguity. There are several factors responsible for stimulating challenges in a VUCA environment. At the same time, these very factors can be responsible for becoming the game-changers in a VUCA world by identifying, prioritizing, analyzing, and deducing the challenges or threats, also known as the IPAD model.

   There are several other ways of eliminating and countering the challenges of a VUCA world, such as the OODA model, popularly known as the OODA loop. The VUCA prime model is a classic method to overcome the challenges of VUCA by converting them into vision, understanding, clarity, and agility.

1.4 RESEARCH GAP

While analyzing the secondary data on the game changers in a VUCA world, the significance of data, technology, and organizational leadership is provided in great detail with many meaningful insights and inferences. However, it has been observed that the significance of policy framing and implementation in a VUCA world remains untapped.

1.5 PROBLEM STATEMENT

As mankind and society traverse the fourth phase of the industrial revolution, they continuously encounter threats in the form of VUCA. The concept of VUCA has been in the limelight since the beginning of evolution and innovations. As society progresses and develops, these threats continue to increase. The acronym VUCA has expanded, leading to the emergence of the concept of "VUCAFU," where the last two alphabets refer to the 'fear of the unknown' and 'unprecedentedness.' There are multiple factors responsible for the rise of the VUCAFU world. However, if these factors work together, they can help society counter the challenges and become the game changers in a VUCAFU world.

The objective of this chapter is:

- The objective is to examine the interrelationships between crucial factors that contributed to the emergence of a VUCA world.
- How do these factors serve as game changers in a VUCA world, thereby establishing a sustainable environment?
- Explore effective strategies for mitigating the threats posed by VUCA.

1.6 RESEARCH METHODOLOGY

The chapter is written after analyzing the secondary data and literature available on the game changers in a VUCA world. Secondary data is defined as previously available data. The study conducted is descriptive and qualitative in nature. Qualitative research employs methodologies like participant observation or case studies to produce a narrative, descriptive account of a location or practice (Physiopedia, n.d.).

1.7 THE GAME CHANGERS IN A VUCA WORLD

1.7.1 ALLIANCES AND CONNECTIVITY

When it comes to dealing with a VUCA world, inter-governmental, and international alliances play a major role in policy formulation and implementation. Today, countries have focused on increasing their adaptability and agility to ensure that they not only survive but thrive in the VUCA environment. Countries have been forced to move out of their comfort zones and be more open to and integrated with the rest of the world, thereby creating an increase in the formation of strategic, political, and military alliances among countries. There are two main causes for the formation of global alliances:

1. **Connectivity Alliances:** Ever since the emergence of liberalization, privatization, and globalization, as well as the 17 SDGs of the UN, the concept of connectivity has been gaining significance. To sustain and thrive in a VUCA world, connectivity is essential. Connectivity encompasses both physical and digital connectivity. It helps facilitate trade, promote entrepreneurship, develop physical and digital infrastructure, and optimally utilize society's scarce resources. This, in turn, brings about sustainable economic growth. However, the true potential of connectivity remains untapped. Connectivity plays a crucial role in helping society achieve the 17 SDGs (Figure 1.1).

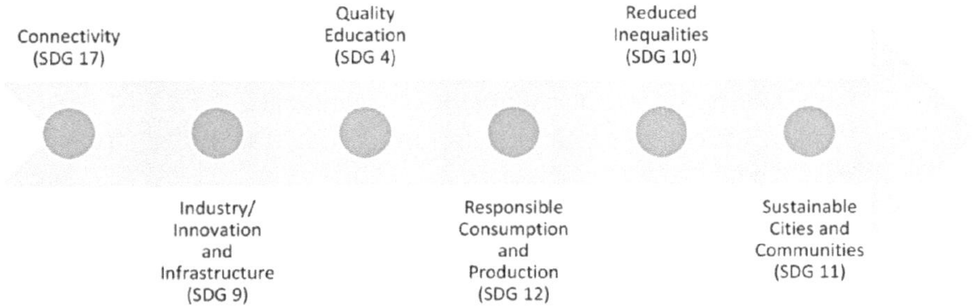

FIGURE 1.1 How connectivity plays a crucial role in achieving SDGs?

Source: Developed by the Authors.

The COVID-19 pandemic and subsequent global lockdown have highlighted the importance of both physical and digital connectivity, as well as the shortcomings in the current connectivity framework. Nevertheless, enhancing connectivity among different global economies would promote industry, innovation, and infrastructure (SDG 9). This, in turn, would create better educational opportunities for students and young people (SDG 4) and lead to improved employment prospects and increased entrepreneurship (SDG 8). By improving the quality of education, job opportunities, and fostering entrepreneurship, we can reduce inequalities within and between economies (SDG 10). Consequently, this would encourage responsible production and consumption of goods and services (SDG 12), ultimately contributing to the development of sustainable cities and communities (SDG 11).

2. **Security Alliances:** A military or security alliance is a coalition of states that collaborate to protect all members from external threats. The application of economics to the study of alliances is driven by three main factors: production, costs, and the theory of public goods. Production is motivated by the fact that national security relies on member nations' contributions, such as weapons and military personnel. Additionally, cost-sharing among allies significantly impacts the provision of optimal security. Lastly, if defense or deterrence is considered a public good, various hypothesis can be derived regarding the organizational structure of alliances, whether they are evolving or already established (Carlos Seiglie, n.d.).

   International alliances such as the UN, WTO, G7, G20, etc., play a crucial role in a dynamic VUCA world. These alliances are responsible for the economic upliftment of society as they trigger policy framing and implementation at a global level. This, in turn, assists national and regional governments in policy framing and implementation in their respective countries or regions. These policies ensure lower living costs, better standards of living, and equality among all sections of society, leading to economic upliftment (Figure 1.2).

1.7.2 ENTREPRENEURSHIP

An entrepreneur is someone who formulates ideas for conducting business, organizes the necessary resources, and integrates them to carry out business activities. Entrepreneurship not only creates employment opportunities but also contributes to innovation, sustainability, and economic development.

Throughout the years, entrepreneurs have played a crucial role in society. As visionaries, they have the ability to turn the weaknesses of the economy

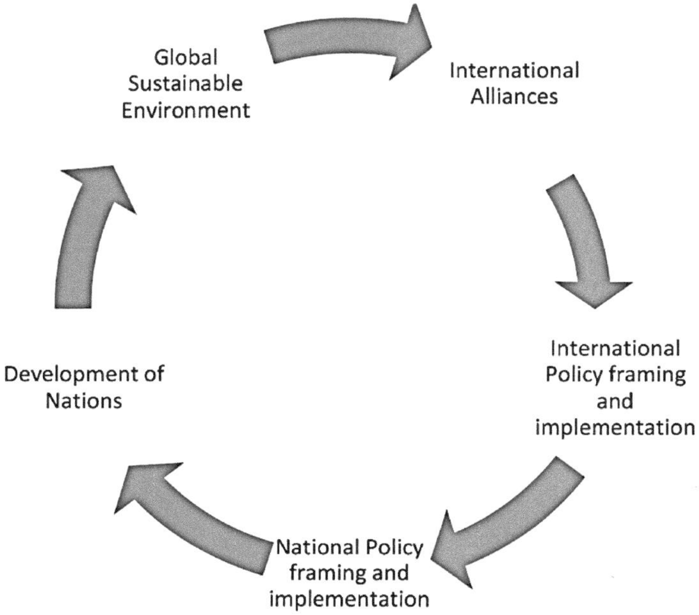

FIGURE 1.2 How do alliances play a crucial role in building a sustainable environment?
Source: Developed by the Authors.

into strengths and transform threats into opportunities for society using their innovative powers. Lionel Robbins described resource scarcity as a major challenge for every economy; however, entrepreneurs must optimally utilize these scarce resources to meet the endless human needs. Liberalization, privatization, and globalization have created vast opportunities for modern entrepreneurs.

Today, an entrepreneur is not only a visionary whose aim is to maximize profits or sales. It is a much broader concept. It encompasses social entrepreneurs, environmental entrepreneurs, technopreneurs, cyberpreneurs, researcher entrepreneurs, infopreneurs, and innovative entrepreneurs who collaborate to promote economic upliftment. Before discussing how entrepreneurship can contribute to sustainability in a VUCA world, we must analyze the threats posed by the VUCA world in this context (Figure 1.3):

1. **Employment Shortages:** Due to cut-throat competition and population explosion, there is a scarcity of employment opportunities. Each company can only hire a limited number of individuals, and their goal is to select the best candidates from the available pool.

2. **Expansion of Red Oceans:** Red oceans refer to markets where numerous firms compete for a larger market share. An increase in the number of firms in the red ocean can intensify competition and lead to unfair trade practices.

3. **Resource Scarcity:** Resources continue to deplete over time. As resources become scarce, it is crucial to ensure their proper utilization in order to create innovative products.

Volatility	Rapid Environmental Changes + Economic Fluctuations + Emergence of New Competitors or substitutes with new strategies
Uncertainty	Acceptance of commodities by society + Emergence of new government policies + Future prospects of the commodity offered in the market
Complexity	Long chain of command + Large number of middle-men + Coordination among various units and branches for global businesses
Ambiguity	Lack of information + Competitor Strategies + Information overload

FIGURE 1.3 Challenges faced by businesses and entrepreneurs in a VUCA world.

Source: Developed by the Authors.

1.7.2.1 ROLE OF ENTREPRENEURS IN A VUCA WORLD

1. **The Creation of Blue Oceans:** Entrepreneurs are visionaries and are expected to possess innovative skills. In a VUCA world, entrepreneurs play a vital role in creating blue ocean strategies by tapping into untapped markets or ideas.

2. **Economic Growth:** Every nation across the globe competes for a higher growth rate. Entrepreneurs use their creativity, imagination, and innovative abilities to contribute to the process of economic growth.

3. **Innovation:** Entrepreneurs play a crucial role in innovating new products or services to improve people's lives. These innovations are responsible for changes in lifestyle, customs, habits, and approaches. Entrepreneurs have a significant impact on shaping the future of mankind and society as a whole.

4. **Converting the Threats of VUCA into Opportunities:** Instead of competing with each other, entrepreneurs should work together for the greater good of society. Infopreneurs, technopreneurs, social entrepreneurs, and all other types of entrepreneurs should collaborate and utilize their combined innovative powers and skills to eliminate the threats posed by a VUCA world.

Rather than viewing the pandemic as a threat to society, entrepreneurs should embrace this challenge and convert it into an opportunity to assist mankind and society in innovation, rethinking, and reshaping their lifestyles. As people's lifestyles and work culture change, taste and preference patterns, market demand, and supply would need to be adjusted accordingly. Thus, there is a vast scope for all categories of entrepreneurship in the new normal.

Figure 1.4 illustrates a scenario in which different categories of entrepreneurs collaborate. Infopreneurs gather information and distribute it to various groups of entrepreneurs. These entrepreneurs then analyze the information and develop or customize products and services according to the needs of the public.

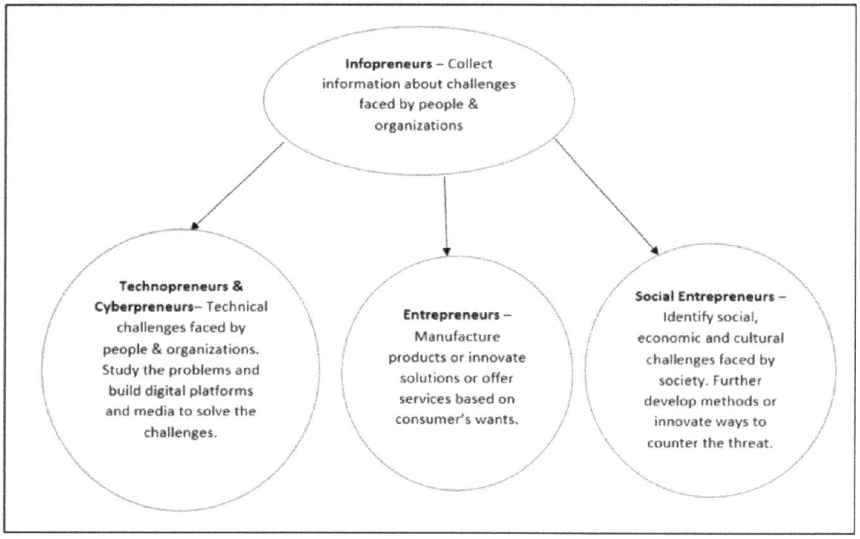

FIGURE 1.4 An integrated working model of different types of entrepreneurs.

Source: Developed by the Authors.

The global pandemic, followed by successive lockdowns, has compelled individuals and society to adapt to a new lifestyle and work culture. This has significantly expanded the opportunities for various categories of entrepreneurs. The emergence of telework culture and digital nomadism has become a new trend not only in western or developed countries but also in developing countries. As people embrace digital nomadism or telework culture, it has also impacted consumption patterns and market demand for goods. Consequently, this has influenced the market demand and supply of goods and services, leading to changes in the socio-economic status of economies worldwide.

Entrepreneurs worldwide play a crucial role in helping nations achieve the 17 SDGs. The VUCA world presents numerous challenges to nations, and entrepreneurs have a pivotal role in addressing these challenges (Table 1.1).

TABLE 1.1 Role of Entrepreneurs in Achieving SDGs

Category of Entrepreneur	SDG
Social entrepreneur	SDG 1, 2, 5, 6, 7, 13, 14, 15, 16, and 17
Technopreneurs	SDG 8, and 9
Edupreneurs	SDG 4
Entrepreneurs	SDG 8, 9, 10, 11, and 12
Healthpreneur	SDG 3
Infopreneurs, cyberpreneurs, others	They may or may not directly contribute to the SDGs but also assist the other categories of entrepreneurs to achieve the SDGs.

Source: Developed by the Authors.

1.7.3 EDUCATION

"Education is the manifestation of the perfection already in man" – Swami Vivekananda (VIVEKAVANI, 2020). Every sector in this VUCA world undergoes constant transformations. Education is no exception. To keep pace with the times, education has shifted its perception from a redundant means of acquiring knowledge to an ever-advancing, technology-incorporated tool. Education and learning have taken the form of mobile applications, massive open and distant learning, and digital textbooks, all available with just one click. Educational institutions have also begun to emphasize the development of a knowledge economy, which includes implementing educational research activities and fostering scientific collaboration among learners.

Preparing children for future outcomes is a vague concept, as one cannot predict the future or what will be required for individual or societal development and sustainability. Additionally, educational institutions cannot truly achieve instrumental goals.

The diagram in Figure 1.5 illustrates the cyclical process of changes in the education system and how it leads to a metamorphosis of society as a whole. One of the most undesirable obstacles that students face is competition. Achieving the highest grades or placements and being the top-ranked student in the class have become highly competitive activities in a student's life. The depth of education is increasing with a vast curriculum, projects, assignments, and internships. This competitive environment poses another threat to students in schools, colleges, and higher learning institutes in a VUCA (volatile, uncertain, complex, and ambiguous) world. Changes in the academic structure and syllabus, along with modernized teaching and learning methodologies, in a robust competitive environment, trigger a chain reaction that compels society as a whole to undergo metamorphosis. This leads to industrialization and increased reliance on artificial intelligence (AI) and machine learning (ML), resulting in technological advancements. Therefore, it is crucial for individuals to develop their skills and keep up with ever-changing technology, leading to socio-cultural change and economic growth. The 2019 pandemic has further increased mankind's dependence on remote learning and promoted the culture of e-learning among students.

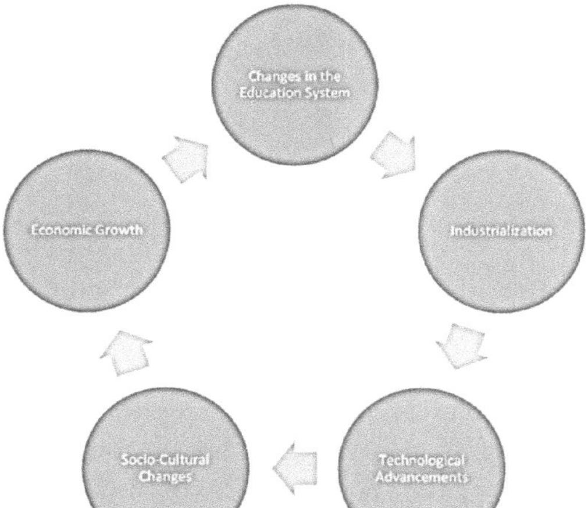

FIGURE 1.5 Cycle process of change in the education system.

Source: Developed by the Authors.

The most effective way to prepare students for a VUCA environment is to focus on the present rather than the future. To ensure that individuals are well-equipped for tomorrow, it is crucial to concentrate on the students in front of us today – their abilities, unique perspectives, and interests. By creating an environment where students feel valued, supported, and acknowledged, as well as providing opportunities for them to apply their skills in creatively solving problems that matter to them today, we can prepare them to bring their full range of abilities, curiosity, creativity, and critical thinking to future VUCA situations that we cannot even anticipate.

1.7.4 TECHNOLOGY

In this era driven by technology, innovations are occurring at an accelerated pace. Advancements such as touch screen technology, cloud computing, the internet, social media, and the widespread use of personal computers in society have led to technological innovation happening much faster than in the past. Consequently, there is now rapid and extensive access to a wealth of information. For instance, the Financial Times made all of its Brexit-related articles freely accessible in the week leading up to the vote, using the phrase 'A Star is Torn' to resonate with the large number of potential new readers seeking information. Despite forecasting a 5% decrease in Q2, 2016 revenues following the Brexit vote, EasyJet launched a new campaign called "Why Not?" with a focus on its mainland European destinations, anticipating a potential surge in people looking to visit those countries for vacations the following summer before any changes to UK citizens' travel rights were implemented (Robinson, 2017).

Technology is a significant catalyst for change in a VUCA world. It has the potential to both amplify and alleviate the challenges that society and mankind face in such a volatile, uncertain, complex, and ambiguous environment. The ways in which technology can support society in navigating the challenges of a VUCA world are as follows:

1. **Artificial Intelligence (AI) and Machine Learning (ML):** In the realms of economics and business, the pace of change is often rapid, characterized by unpredictable fluctuations and turbulence. As the market becomes more volatile, the rate of change accelerates. While AI brings about disruptive effects, it also enables more effective change management. The ability to forecast the future is directly proportional to the level of ambiguity in a given situation. AI, with its superior capacity to analyze data swiftly and accurately, helps to minimize uncertainty. As the number and complexity of components involved, as well as their interrelationships, increase, the analysis becomes more intricate and challenging. However, ML significantly aids this process by identifying systemic relationships. In situations where evidence is insufficient or contradictory, understanding how

to interpret the scenario becomes a challenge. The hazier the situation, the more difficult it is to derive meaning from it. ML and AI data can provide much-needed clarity in such instances.

2. **Big Data and Data Mining:** In the volatile, uncertain, complex, and ambiguous VUCA world, decision-making needs to be more authentic, transparent, and qualitative. Data serves as the lifeblood of every business, and industry decisions lay the groundwork for the future. With data being limitless, various sources provide access to data acquisition. In the VUCA world, we must harness BIG Data for decision-making, whether it is structured, unstructured, or semi-structured. There are 7 V's that characterize big data, and these 7 V's can help mitigate VUCA. They are Volume, Variety, Veracity, Velocity, Variability, Visualization, and Volatility (Sangeet, 2018).

3. **Science Fiction:** What was once science fiction yesterday has become scientific fact today. The science fiction genre holds the potential to benefit businesses and revolutionize organizations. It allows the world to envision itself in a new perspective, one that belongs to the future. Science fiction is a form of predictive fiction that showcases the potential direction of the world. In recent years, many corporations have turned to science fiction for fresh ideas, products, and production methods. Science fiction is valuable not only because it predicts the future but also because it reframes our perception of the world we inhabit. Companies like Google, Microsoft, and Apple have even enlisted science fiction writers as consultants to infuse elements of the fictional world into our reality using technology.

   People contemplate the future and its mysteries, and these dreams have given rise to speculations. These speculations, in turn, have transformed into inventions with the aid of technology. This cycle perpetuates in an endless loop, commonly referred to as progress, which ultimately leads to sustainable development.

1.8 STRATEGIES TO ADDRESS THE CHALLENGES OF VUCA

1.8.1 THE IPAD (IDENTIFY, PRIORITIZE, ANALYZE, AND DEDUCE) MODEL

VUCA reflects the reality in which society resides. Society is increasingly unstable and subject to rapid change. It signifies that unforeseen events can occur at any time, and more frequently than in the past. VUCA has become the new norm, and society must be equipped to navigate it on a daily basis in order to ensure long-term sustainability. The challenge lies in the fact that, despite living in this reality, the majority of leaders are ill-prepared to tackle it.

The IPAD model, consisting of four stages, offers a four-step approach to address the challenges posed by the VUCA world. These stages are as follows (Figure 1.6):

1. **Identify:** Conquering a challenge or threat begins with its identification. This initial step is crucial in addressing any shortcomings. Environmental scanning is essential for identifying threats in the environment. It is not only important for businesses and entrepreneurs, but also for other sectors such as international relations and politics. Thorough environmental scanning can help identify the threats that a VUCA world presents to society.
2. **Prioritize:** Once all the threats have been identified, it is essential to prioritize these threats based on their urgency and impact. Threats that could have a severe effect on society and the economy must be prioritized over the other less important threats.
3. **Analyze:** Once the threats are prioritized, the prioritized threats are then carefully analyzed and studied from various angles by experts. It is essential to carefully analyze the threat to understand its impact on society.
4. **Deduce:** After analyzing the threats, the experts conclude to try and either eliminate the threat or convert it into an opportunity or at least work to reduce the impact of the threat upon society.

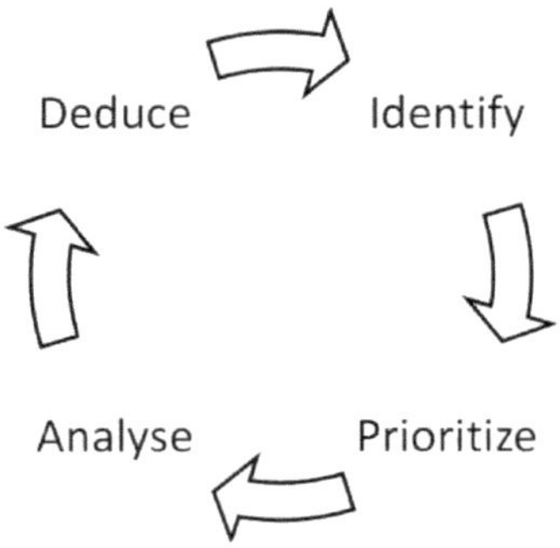

FIGURE 1.6 IPAD cycle.

Source: Developed by the Authors.

These four stages of the IPAD model continue in an endless loop to assist society in overcoming the challenges of a VUCA world. Each threat or challenge must first be identified and ultimately resolved.

The VUCA environment is characterized by turbulence. It emphasizes the need for extensive knowledge and the application of new and practical approaches. An organization cannot avoid the complexities of the VUCA world, but it can learn to adapt and thrive. The IPAD model provides an alternative approach to dealing with the components of VUCA. Its foundational principles are similar to the existing VUCA Prime model, with only surface elements differing. It is simple and relevant in nature. Undoubtedly, more models like this will be developed in the future. The sole purpose of these models should be to analyze each variable of the environment through the lens of the VUCA world and strategize accordingly.

1.8.2 OTHER SOLUTION MODELS OF VUCA

The 2019 pandemic is not only a significant healthcare issue impacting governments and people worldwide, but it is also wreaking havoc on businesses and society. The coronavirus pandemic has had a detrimental impact on nearly every industry. Unemployment rates are at an all-time high, and the rate of GDP growth is highly unpredictable. Companies have gone bankrupt, and factories have ceased operations. Even service industries, particularly the IT industry, have embraced the telework culture, leading to management and efficiency challenges. This new paradigm, brought about by the epidemic, has revived Darwin's principle of survival of the fittest, where the key to seizing new opportunities lies in swiftly implementing and embracing changes in business models. Many severely affected sectors require a fresh and innovative framework to be established within traditional society, enabling them to regain momentum.

Various models offer strategies to overcome the challenges of a VUCA world. Some of them include:

1. **VUCA Prime:** Introduced by Robert Johansen in 2007. According to this model, each VUCA challenge can be addressed with a specific positive response. Vision supersedes volatility, emphasizing the need for a comprehensive and compelling vision of a sustainable future and the steps to achieve it. Understanding mitigates uncertainty, urging society to explore the various factors and sources contributing to uncertainty. Clarity reduces complexity, encouraging a careful

analysis of complex situations to gain a clearer understanding and draw conclusions—agility, counters ambiguity, emphasizing the importance of quick thinking and decisive action to eliminate ambiguity.

2. **OODA Loop:** VUCA describes the modern environment as one filled with VUCA. Making intelligent decisions is the key to survival and success in this environment. The OODA loop is an interactive and iterative process that involves repeating the cycle, measuring and observing results, assessing and amending the initial choice, and moving on to the next stage. The first stage in the OODA Loop is observation. It involves identifying threats and gaining a general understanding of the internal and external environments. The next stage is orientation. During the orientation phase, one must reflect on the observations made and decide on the next course of action. Making a conscious decision requires a high level of situational awareness and comprehension. This is followed by the decision phase, where recommendations are made for an action or response plan, taking into account all possible outcomes. The final stage is the action stage. Action refers to implementing the decision and making any necessary adjustments as a result of the decision.

   Considering the events occurring worldwide, it can be confidently stated that we are operating in a VUCA world. The acronym 'VUCA' was initially coined at the United States Army War College, but today it is widely used to describe the challenges faced by society in general and the business sector in particular.

1.9 CONCLUSION

The challenges and threats posed by a VUCA world are akin to the ceaseless tides of the sea. As society advances and innovations unfold, these challenges and threats persistently escalate. However, within a VUCA world, there exist numerous game-changers. These game-changers are intricately interconnected and directly or indirectly contribute to society's pursuit of the 17 SDGs established by the UN in 2016.

Volatile, uncertain, complex, and ambiguous (VUCA) can be perceived as a method of enhancing leadership and management competencies. Hazards can be transformed into advantages if society places emphasis on domains such as decision-making, innovation, implementation, and creativity. Being slow-moving and outdated is not a viable choice in this VUCA world. Novel, pragmatic, and unconventional approaches are necessary to effectively navigate the VUCA environment.

KEYWORDS

- **entrepreneurship**
- **identify**
- **prioritize**
- **analyze**
- **deduce**
- **IPAD model**
- **OODA loop**
- **sustainable development goals**
- **volatility**
- **uncertainty**
- **complexity**
- **ambiguity**
- **VUCA prime**
- **VUCAFU**

REFERENCES

Anand, J., (2020). *Connectivity in an Age of Volatility, Uncertainty, Complexity and Ambiguity (VUCA): Realizing SDG9 Amid the COVID Pandemic.* Retrieved from: pathfinderfoundation.lk:https://pathfinderfoundation.lk/images/pdf/Conference_Materials/SOUTH_AFRICA_-_ENHANCING_CONNECTIVITY_SDG_9_and_COVID.pdf (accessed on 09 March 2024).

Carlos, S. S. M., (n.d.). *Global and Regional Security Alliances*. Retrieved from: eolss.net: https://www.eolss.net/sample-chapters/c13/E6-28A-04-03.pdf (accessed on 09 March 2024).

Elena, S. D. S., (2020). Strategies for developing competencies for the VUCA world. *Proceedings of the 3rd International Conference Spatial Development of Territories* (Vol. 81, pp. 362–366). Advances in Economics, Business and Management Research.

Gianfagna, M., (2021). *What is Moore's Law?* Retrieved from: synopsys.com: https://www.synopsys.com/glossary/what-is-moores-law.html#:~:text=Definition,as%20E%20%3D%20mc2 (accessed on 09 March 2024).

Korsakova, T. V., (2020). *Higher Education in VUCA-World: New Metaphor of University.* Retrieved from: revistia.org: https://revistia.org/files/articles/ejis_v6_i1_20/Korsakova.pdf (accessed on 09 March 2024).

Lorraine, E. C. F., (2021). *Evidence-Based Policymaking in a VUCA World.* Retrieved from: unctad.org: https://unctad.org/system/files/official-document/diaeia2021d3a8_en.pdf (accessed on 09 March 2024).

Moreno, O. M., (2020). *Coordinated Governance in the VUCA Scenario.* Retrieved from: hrcak.srce.hr: https://hrcak.srce.hr/file/384439 (accessed on 09 March 2024).

Nathan, B. J. L., (2014). *What a Difference a Word Makes: Understanding Threats to Performance in a VUCA World.* Retrieved from: researchgate.net: https://www.researchgate.net/profile/Nathan-Bennett-5/publication/263926940_What_VUCA_really_means_for_you/links/5a53813e458515e7b72eb049/What-VUCA-really-means-for-you.pdf (accessed on 09 March 2024).

Physiopedia, (n.d.). *Qualitative Research* – Physiopedia. Retrieved from: physio-pedia.com: https://www.physio-pedia.com/Qualitative_Research#:~:text=Qualitative%20research%20is%20research%20using,a%20form%20of%20interpretive%20sociology (accessed on 09 March 2024).

Robinson, R., (2017). *Technology in the VUCA World.* Retrieved from: medium.com: https://medium.com/@redronrich/technology-in-the-vuca-world-c71d6c8ee553 (accessed on 09 March 2024).

Sangeet, V. P. S., (2018). Big data – New trend of change in complex corporate world. *Globus An International Journal of Management & IT*, 4–6.

Suryanarayana, P. A., (2022). *Insights Into Leadership Mandates and Imperatives in a VUCA World.* Retrieved from: https://www.journalppw.com/index.php/jpsp/article/view/4578 (accessed on 01 Aptil 2024).

Vivekavani, (2020). *Education is the Manifestation of the Perfection Already in Man.* Retrieved from: https://vivekavani.com/education-manifestation-perfection-man/ (accessed on 09 March 2024).

CHAPTER 2

Socio-Intercultural Sustainable Natural Resources Management and Governance

JOSÉ G. VARGAS-HERNÁNDEZ[1] and OMAR C. VARGAS-GONZÁLEZ[2]

[1]*Posgraduate and Research Division, Instituto Tecnológico Mario Molina, Zapopan, Mexico*

[2]*Department of Systems, National Technological Institute of Mexico, Mexico*

ABSTRACT

This chapter aims to analyze the interrelationships and implications of socio-intercultural sustainable natural resources management and governance. It is based on the assumption that natural resources management and governance should be determined by socio-intercultural sustainability. The method used is a reflective analysis, which includes a literature review and an analysis of sustainability trends related to these issues.

2.1 INTRODUCTION

The management and governance of natural resources is an area that fosters innovation and experimentation. Governance encompasses the relationships between individuals, power structures, corporate management, and other critical factors that influence performance. Natural resources governance and management involve activities related to agroecology, forestry, water allocation, and tourism, all supported by biodiversity conservation, the preservation of natural resources, and land-scale management strategies.

Contemporary Digital Transformation and Organizational Effectiveness in Business 4.0.
Nilanjan Ray, Sudin Bag, & Tapas Kumar Chatterjee (Eds.)

Natural resources management represents a contemporary approach to governance that integrates various actors and stakeholders in collective action. This approach aims to foster the creation of knowledge, learning, and experimentation, all with the goal of improving conditions (Head & Ryan, 2004; Morrison et al., 2004). Socio-interculturality varies in its pursuit of equity, achieved through the integration of traditional indigenous ecological knowledge and modern scientific knowledge within the dimension of environmental management governance in societies and nation-states. Traditional knowledge in environmental management influences the integration of environmental governance and the management of socio-ecosystems.

The new governance perspective addresses the problems of uncertainty and complexity in the management of new natural resources. It takes into account the interdependencies and connectedness among diverse stakeholders and actors at different scales and jurisdictions. The new governance approach is a mode of governing that aims to cope with complexity and uncertainty, emphasizing learning and experimentation processes (Lenton, 2002). It efficiently manages resources, knowledge, and skills compared to conventional government approaches.

Local people generate and develop usable knowledge and skills regarding outcomes, changes, and impacts. This strengthens governance and creates potential for research. It also strengthens the role of a socio-intercultural governance committee, integrating local indigenous and scientific knowledge into traditional governance of planning and innovative adaptive management that embodies the principles of co-motion and co-production (Howitt et al., 2013; Muller, 2014).

The reflective analysis of this study is based on socio-intercultural sustainable natural resources management and governance. To achieve the aims, the analysis begins by describing and explaining the sustainability and socio-intercultural natural resources management issue. Then, the analysis focuses on the issues related to the sustainable governance of natural resources and sustainable natural resources management. Finally, a discussion and some concluding remarks are provided.

2.2 SUSTAINABILITY AND SOCIO-INTERCULTURAL NATURAL RESOURCES MANAGEMENT

A sustainable socio-ecological system maintains a cycle of pressures and disturbance events, ensuring its capacity to endure through its characteristic diversity of major functional groups, processes, services, and utility (Dawson et al., 2010, p. 2845). The combination of indigenous ecological knowledge and modern science is influenced by adaptive management and governance systems of natural resources and the environmental context. The utilization of sustainable natural resources arises from the uncertainty of unintended consequences of activities within the complexity of the ecosystem.

Sustainable socio-intercultural natural resource management systems exist where natural resources coexist with diverse local interests. Socio-intercultural natural resource governance and management, in terms of intangible assets communication and measurement, represent socio-intercultural capacity deficits of local communities and indigenous people, who are confronted with the challenges of finding innovative solutions (Howitt et al., 2013; Muller, 2014).

Socio-intercultural governance has become a concern among academics studying indigenous people worldwide since the First Indigenous Governance Conference held in 2002. Indigenous people have developed equitable socio-intercultural partnerships in land and sea management processes and planning techniques (Ansell & Koenig, 2011; Ens et al., 2012a, b; Hoffmann et al., 2012; Australian Government, 2013; Hill et al., 2013; Jackson et al., 2015).

The socio-intercultural governance of indigenous management partnerships requires processes and tools to develop an innovative and adaptable impact on land and sea. These partnerships can integrate scientific and indigenous community knowledge to embody co-production principles of adaptive management. They can be established to create innovative, socio-intercultural, and interdisciplinary bodies and committees formed through collaborations adopting multiple evidence-based approaches.

Socio-intercultural natural resource systems coexist with natural resource management and co-management systems aimed at addressing socio-intercultural competences to resolve claims from local communities and the state. Co-management systems emerged in the 1990s, cutting across cultural differences and purposes (Natcher et al., 2005; Berkes, 2009). However, cultural differences might hinder collaboration between organizations due to different cultural philosophies, values, management styles, and mechanisms of governance.

The exercise of national state power can sometimes undermine the responsibilities and values of indigenous and local governance and may not always be open to co-management or partnership arrangements (Agius et al., 2002; Sobeck & Angius, 2007). The state power has the ability to utilize political, legal, public policy, and financial resources to promote good governance at the local level. However, it is important to recognize that regional and local governance should not be limited to technical management tasks alone. It should also embrace local knowledge and expertise systems, while respecting local self-determination and authority in co-management arrangements that facilitate socio-intercultural governance.

Socio-intercultural natural resources management systems are characterized by ontological pluralism, which acknowledges and incorporates the diverse human-nature relationships (Howitt & Suchet-Pearson, 2003, 2006). This ontological pluralism in socio-intercultural natural resources management recognizes and values indigenous knowledge and governance on par with scientific, economic, and democratic assumptions (Wohling, 2009; Turnhout, 2010). A collaborative approach to governance entails the active involvement of all governing actors and stakeholders, with access to multiple sources of information and natural resources at various levels, while appreciating diversity within multilevel systems.

The capacity building of socio-intercultural natural resource management systems is influenced by the interaction between ontological pluralism and various factors including biophysical, economic, socio-cultural, and political factors (Howitt et al., 2013).

Frameworks for socio-intercultural natural resource management systems require collaboration and local administration with indigenous institutional governance to ensure cultural outcomes while challenging dominant cultural constructions (Weir, 2009; Muller, 2012; Wohling, 2009). Collaboration refers to the co-management processes between local communities, indigenous institutions, and dominant institutions of the nation-state (Muller, 2008c). Collaboration necessitates socio-intercultural competences and capabilities in natural resource management systems.

New governance theory repositions the components of governance within institutional and organizational structures, placing them under the authority of private actors who possess the necessary capabilities and resources. This approach aims to alleviate the burden on the state while enabling more effective exercise of governance. Institutional decentralization involves the transfer of powers and resources from the central government to local government institutions and actors (Agrawal & Ribot, 1999).

Natural resource management governance requires national, local, and community institutions to be more democratic, decentralized, and mutually supportive (Agrawal & Ribot, 1999). Nation-state government institutions and agencies lack the socio-intercultural capacities to effectively address the needs of local communities and indigenous groups involved in natural resources management systems Governance authorities must have access to all available sources of knowledge and resources in order to make informed decisions and develop effective solutions for the design and implementation of a decentralized governance system for natural resources management. This can be achieved through adopting a collaborative approach and avoiding top-down or vertical governance structures.

The implementation of a sustainable and systemic resolution for competing claims between local communities and the state in socio-intercultural nature resource management systems requires addressing various challenges, primarily capacity deficits. These deficits, which exist at the individual, community, agency, and institutional levels, limit the operational effectiveness and socio-intercultural communication within natural resources management systems. Collaborative management structures and processes that align with dominant human-nature relationships can help address these socio-intercultural capacity deficits (Suchet-Pearson & Howitt, 2006).

Organizational administration and governance issues are often linked to loosely defined constitutions, organizational structures, and structural problems stemming from perceived poor management, mismanagement, and lack of good governance.

The management and utilization of sustainable natural resources necessitate normative support and call for new governance principles and arrangements that must be adapted to specific situations, even if they were originally developed for different contexts. However, applying governance principles developed at other levels of governance, such as national, regional, or local scales, to the governance of sustainable natural resources may prove more challenging.

Socio-intercultural systems for managing sustainable natural resources should be a rigorous and coherent discipline, accessible to local and global economic, social, political, and environmental processes. They should also be responsive and accountable to the individuals, institutions, and groups involved at different levels of governance. According to the authors (Howitt et al., 2013), the new geographies of coexistence have an impact on the capacity deficits of processes, institutions, and knowledge systems, resulting in numerous systemic failures that affect the socio-intercultural capacity of socio-intercultural natural resource management systems.

Poverty, exclusion, and marginality are some of the structural limitations to achieving sustainable management and governance of natural resources. This can be addressed through social learning and communicative action processes that bring together and coordinate activities among stakeholders and actors involved. The involvement of all stakeholders is crucial for the success of this initiative, as it promotes shared understandings and cooperation through the creation of a platform.

Strategic action to address governance constraints and limitations involves negotiation between stakeholders and actors, aiming to achieve more sustainable management and governance of natural resources. Social learning processes play a significant role in shaping sustainable management and governance practices and the state of natural resource systems. This requires the development of new theoretical frameworks to guide assessments and monitoring approaches.

2.3 SUSTAINABLE GOVERNANCE OF NATURAL RESOURCES

Organizational flexibility, leadership, sustainable governance, and institutional strength are interconnected components of a sustainable organizational and community ecology system. This system is capable of creating and maintaining cultural identity, as well as family and community relationships, while adhering to organizational governance standards and financial accountability. Extended forms of families have a governance structure embedded within layers of regional and community governance. However, the perception of family relationships and participation in community governance and organizational management tends to be negative. This is partly due to the association of families with different forms of corruption, self-interested decision-making, nepotism, violence, disputation, functionalism, debilitating factionalism, and other dysfunctionalities.

The issue of governance has gained increasing familiarity since the 1990s in relation to organizational management. It has evolved to address capacity building and solve deficiencies. However, the term governance encompasses a wide range of meanings associated with different variables and their effectiveness.

The concept of socio-intercultural sustainable leadership is emerging as a counterpoint to the dominance of classical leadership theories and practices. While planning and design alone cannot guarantee a creative, socio-intercultural, and sustainable city, strategic objectives for urban interculture can be established (Clemente, 2008), allowing for the emergence of socio-intercultural sustainable leadership. The leadership identity of any local community's sustainable socio-intercultural governance is formed through a negotiated process involving all stakeholders within a politically charged organizational and social context (Sinclair, 2007, p. 128). Design principles of socio-intercultural governance can address and support initiatives related to community autonomy, legitimacy, power, representation, and accountability.

This prevailing perspective on socio-intercultural governance in natural resource management should be challenged by combining tools from indigenous communities and urban planning agencies (Lunkapis, 2010; Howitt & Lunkapis, 2010). The various layers of governance can be monopolized by small elites or "big men" (Hughes, 2005, p. 16) who exploit the benefits by controlling natural resources in aboriginal local communities. Community mapping and social planning systems, supported by an analysis of the impact on the assessment framework of values, rights, and experiences in natural resources management, make decisions based on public participation, self-determination, and self-governance (Loomis, 2000; Ridder & Pahl-Wostl, 2005; Xanthaki, 2007).

Indigenous people ensure their right to preserve their traditional system of governance, secure their natural resource caring and management, and have certainty of the patterns of trade and movement as the owners of the territory where these issues occur (Howitt, 2012). Indigenous and local authorities struggle to achieve recognition and autonomy in natural resources management systems and resource management.

People face pressures from both domains, the local community and mainstream society, to develop the traditional leadership identity enabling intercultural events and articulations to be incorporated and occur to perpetuate a sustainable intercultural governance system. The nature of this socio-intercultural governance system should be based on common assumptions and public understanding to strengthen the sustaining practices and avoid vulnerability (McKenna, 2004).

This system has the potential to develop significant resilience as a legitimate component of effective good governance. However, when the state exerts its power over the community, the transformative benefits of local leadership and governance are greatly diminished. The fields of socio-intercultural governance are deeply intertwined with similarities and differences in socio-cultural forms of thought, personhood, institutions, and actions, leading to mutual exposure of engagements and transformations (Smith, 2007; Gluckman, 1968).

The cognitive component of institutional capital in community sustainable governance is shaped by the identification of community history, relationships, governance, behaviors, values, goals, operations, and other components between community organizations.

Coordinated actions for more sustainable forms of governance of natural resources, based on strategic and communicative action, are related to the material and social resources, as well as the capabilities of the participating social actors. The institutionalized forms of competitiveness of communities and organizations, taking strategic governance roles for the acquisition of scarce natural resources, may result in counterproductive and dysfunctional pressures for more contested governance. Other forms and approaches of governance may contribute to widening or narrowing the public–private gap (Cole, 2011).

Orchestration is a concept rooted in New Local Governance and New Public Management theoretical approaches, shifting away from centralized administration towards private regulation functions and instruments such as market-based mechanisms, competition in the provision of public services, deregulation, outsourcing, etc. (Lane, 2000). The shift from government to governance converges from stakeholders and public politics perspective towards citizens and private management (Bernstein, 2001; Bernstein & Cashore, 2007).

The recommendations for different forms of public and private governance range from institutional coordination, national regulations, and treaties such as PPPs and private arrangements, to multiple authority centers separating private and public governance and widening the engagement gap, decentralizing governance, and difficulty in bridging them (Falkner & Vogler, 2010; Orts, 2011). Some of these sustainable socio-intercultural initiatives are motivated by failures of public governance at the level of public-private partnership interactions (Schäferhoff, Campe, & Kaan, 2009).

Partnerships, strategic alliances, and natural resources can be mobilized across networks by nodal leadership representing knowledge and authority capable of maintaining and reproducing socio-intercultural relationships. A strategic organizational alliance may be formed by a networked governance model structured around a hub to maintain the valued relationships, financial, and management support.

Creating and developing a network to provide the channels for knowledge creation and transfer, facilitate dissemination, and provide assistance to the community is part of good sustainable governance. Deliberate initiatives for long-term socio-intercultural contacts require a network of sustainable communities where residents of diverse backgrounds can meet, developing inclusive social and cultural ties.

The design of governance principles for common property of natural resources institutions sets the starting point at Ostrom's (1990, p. 90); and Borrini-Feyerabend et al. (2006) who developed a code for protected area governance. The Lisbon Principles were developed by experts for the sustainable governance of marine and coastal resources (Costanza et al., 1999).

Communicative action contributes to the development and transfer of technical and managerial knowledge that is useful for more sustainable practices in natural resource management and use.

Some institutions of local governance collapse due to poor policy design and implementation framework management and financial administration, leading to flexible and slippery implementation of sustainable decision-making processes. Community organizations can contribute to debilitating or strengthening internal conflicts, depending on whether they are at odds with the community members and each other, financial management, and governance institutions. A wide range of local governance issues includes local management responsibilities and policymaking, separation of powers and governing roles, democratic representation models, organizational models of regionalization, employment contract conditions, human resource (HR) management, codes of conduct and conflict of interest, community communication, etc.

This situation distances the local governance policy framework from culturally-based processes and systems that acknowledge and determine local government regions and communities. Local governments and the state exercise power and authority over local organizations and individuals, compromising good governance and intensifying struggles over resources, energy, and other values. These struggles for good governance involve the understanding of state and local governments regarding the complex system of values at play in socio-intercultural spaces. Local sustainable socio-intercultural governance lies at the intersection of community vision and aspirations, corporate social responsibility, and economic activities.

2.4 SUSTAINABLE NATURAL RESOURCES MANAGEMENT

Socio-intercultural settings are characterized by the economic, social, and environmental involvement of multiple cultural groups, with new arrangements to mobilize resources in sustainable forms and resolve local claims in line with the different interests of the stakeholders involved. The socio-intercultural sustainable natural resources management systems support a more equitable and sustainable security of coexistence for populations facing the disadvantages of cultural exclusion and marginalization resulting from policy state-dominated frameworks and avoid evaluations leading to conflicts.

Socio-intercultural competence is related to the efficacy and achievement of socio-intercultural systems and sustainable outcomes for local participants and stakeholders, developing a sustainable capacity building of natural resources management. Developing a critical understanding of socio-intercultural capacity building for governance that sustains local and indigenous knowledge is crucial for the tasks of caring for government, society, and culture (Gibbs, 2009; Howitt & Lunkapis, 2010; Weir, 2011). This understanding is essential for socio-intercultural sustainable natural resources management.

Building the socio-intercultural capacity of the natural resource management system is implicated in the coexistence and co-management of local cultures, communities, institutions, and governance, leading to sustainable solutions. Socio-intercultural governance is contested based on ethnographic accounts of the culture of governance in the inter-institutional relationships between indigenous communities and the nation-state, posing questions about the nature of postcolonial indigenous governance.

The systemic governance capacity of indigenous communities in their relationships with government institutions, academic centers, and other resource centers is weak, making it difficult to resist the imposition of national state agendas, structures, and practices. Socio-intercultural governance systems are framed within a postcolonial socio-intercultural context, where the state holds sovereign power and limited local government jurisdictional authority and legislation.

Natural resources management is a contemporary approach to governance that involves integrating diverse actors and stakeholders in collective action to foster knowledge creation, learning, and experimentation aimed at improving conditions (Head & Ryan, 2004; Morrison et al., 2004).

Socio-intercultural sustainable natural resources management systems are characterized by ontological pluralism, which encompasses the diverse human-nature relationships (Howitt & Suchet-Pearson, 2003, 2006). The ontological pluralism of socio-intercultural sustainable natural resources management acknowledges the value of indigenous knowledge and governance in collaborative management, on par with scientific, economic, and democratic assumptions (Wohling, 2009; Turnhout, 2010).

The dialog of socio-intercultural pluralism, open to alterity and complexity in a global perspective, is at the core of good governance. Political and social actors are situated within a social context through interactions and mediating instances of the local polity, based on principles of good governance with an emphasis on hierarchies, boundaries, planning, risk management, and more.

Transnational resource governance schemes go beyond the implementation of rulemaking and overlook the normative multi-stakeholder model beyond the state. They are more globally functional-oriented in the decision-making process and engage with transnational and intergovernmental institutions and organizations. Comparisons of transnational governance systems and processes observe the different levels of legitimacy and effectiveness in design and implementation schemes, such as in transnational natural resource management and governance (Meridian Institute, 2001; Reinicke et al., 2000; Witte et al., 2003).

Governance theory has limitations in existing concepts for the analysis of institutional and cultural challenges posed between relativism and universalism by sustainable global value chains, in order to offer sustainable management methods and instruments. The governance activities must be clearly stated to transparently address any potential conflicts of interest, in order to provide better service delivery functions, sustainable management of indigenous community-owned resources and lands, support economic independence, business opportunities, cultural priorities, and welfare opportunities.

Theoretical frameworks of sustainable global value chain governance have deficiencies in offering more pragmatic, conceptually-oriented perspectives for analyzing organizations operating in socio-intercultural environments. Socio-intercultural dialog requires the sharing of values and visions between diverse cultures and when possible, the establishment of socio-intercultural partnerships to create and utilize multicultural learning spaces and build sustainable societies. The multicultural approach is a static notion of group identity.

Collaborative spaces operating in socio-intercultural sustainable natural resources management systems are based on existing priorities, imperative ethical values, and the ability to work and develop sustainable communicative processes with individuals who have different or contrary frames of reference (Allenby, 2006, p. 8). The socio-intercultural and interpersonal cross-scale dimensions engage individuals in natural resources management to foster collaborative spaces (Suchet-Pearson et al., 2006; Suchet, 1999; Wright et al., 2012).

While cultural intelligence is associated with governance mechanisms and sustainable collaboration, cultural differences impact and influence the relationship between contract governance and sustainable collaboration levels. Contract and relational inter-organizational governance have a positive effect on sustainable collaboration when motivational cultural intelligence is high, particularly in emerging economies. The ability to transition into a socio-intercultural community is linked to organizational governance that supports hybrid local economies.

Socio-intercultural sustainable natural resources management system frameworks require coordinated action and local administration with indigenous institutional governance to safeguard cultural outcomes while challenging dominant cultural constructs (Weir, 2009; Muller, 2012; Wohling, 2009). Governance in Indigenous regions and communities is a critical concern. Indigenous governance, supported by power-sharing with other agencies, enhances the sustainable integration of socio-intercultural knowledge and contributes to the development of robust and effective governance. Indigenous-driven governance and co-governance foster collaborations between indigenous ecological knowledge and modern science for sustainable social-ecological systems (Hill, Grant, George, Robinson, Jackson, & Abel, 2012).

Sound community indigenous governance can be attributed to the maintenance of stability in non-indigenous management bureaucracy and ensuring seamless succession and continuity. In certain cases of identified periods of indigenous community governance tensions, deferring non-indigenous management may result in the abrogation of governance responsibilities, despite the organization experiencing impressive performance and growth. Points of tension may arise between local government officials and community members regarding perceived inequalities in the distribution of natural resources.

The socio-intercultural community governance is characterized by dimensions such as power or de facto sovereignty in decision-making and self-rule, legitimacy, and cultural alignment, access and ownership of natural resources, and accountability and governing institutions. However, these dimensions are adjusted to the socio-intercultural and political contexts of community governance. While community governance arrangements can be distinct, they work together in partnership and in accordance with management responsibilities to promote shared commitment and resilience.

Co-motion requires socio-intercultural competencies and capabilities in managing natural resource systems. Co-motion is associated with co-management processes between the local community, indigenous institutions, and the dominant institutions of the nation-state (Muller, 2008c). The context of community governance and administration is intricately linked to institutional and natural resources in complex ways (Sanders, 2006), which can serve as a foundation for the further development of indigenous community local governance. Political settlements and co-management agreements can result in institutions that safeguard and ensure equitable and sustainable outcomes (Ens & McDonald, 2012; Ens et al., 2012a; Barbor & Schlesinger, 2012).

The socio-cultural sustainable management systems of natural resources provide support for a more equitable and sustainable coexistence for populations facing cultural exclusion and marginalization resulting from state-dominated policy frameworks, thereby avoiding conflicts. The socio-cultural and interpersonal cross-scale dimensions engage individuals in the management of natural resources, fostering the creation of collaborative spaces (Suchet-Pearson et al., 2006; Suchet, 1999; Wright et al., 2012).

Developing a comprehensive understanding of socio-cultural capacity building for governance that supports local and indigenous knowledge is crucial for the sustainable management of natural resources (Gibbs, 2009; Howitt & Lunkapis, 2010; Weir, 2011). This knowledge is essential for self-determination and the sustainable socio-cultural systems related to natural resources. Collaborative efforts are necessary to achieve more sustainable and equitable outcomes (Esteva, 1987). The legal and regulatory system forms the foundation for local indigenous knowledge, which can be integrated through collaborative efforts.

Socio-intercultural spaces allow for coexistence between local indigenous communities and state domains and regulations in territories. Planning agencies can develop socio-intercultural capacities for projects, land use planning, zoning, and resource allocation practices, all subject to planning and development controls. A fair governance framework ensures fairness in decisions on resource allocation, benefiting all stakeholders involved according to an agreed strategic plan.

Government bureaucrats in state systems can manage contingencies through risk management, planning, and control of enclosed spaces. Regular reviews of planning and control processes are part of governance, without causing governance problems due to the complexity of boundaries.

The dominant land use and development planning systems, centered on natural resources management, often adopt a top-down approach that disregards the public interest and undermines any expression of socio-intercultural governance based on local indigenous values. However, indigenous communities have a long-standing tradition of establishing innovative socio-intercultural governance committees. These committees integrate knowledge, provide strategic advice, and make decisions on important matters without undermining their formal authority.

Effective community and organizational governance requires interconnectedness among all involved participants, ensuring integration and coordination across all governance levels and organizations aligned with the plans, activities, and priorities of sustainable natural resources management. International governance organizations have two options to engage in this process: regulatory collaboration and orchestration. Despite limitations such as autonomy constraints, lack of resources and necessary skills, and opposition from national states, these organizations can still play a crucial role.

An instrumental rational governance is founded upon a functional interconnectivity to establish interdependencies across all levels of government, territory, and policy scales (Dovers, 2005). The process of altering the relationship between governance and management has incorporated the status requirements of formal governance. Members of management were required to operate impartially and without vested interests, thereby fulfilling the objectives to benefit the population under the application of a Weberian rationality of impersonal distribution (Gerritsen & Straton, 2007, p. 166).

The domination of western management over aboriginal community participation has been imposed by state control and intervention. Governance and the culture associated with socio-intercultural organization have formed a political alliance between the western-style bureaucracy and aboriginal self-determination, in order to meet the indigenous governance aspirations of the community. This alliance accommodates both western and indigenous social norms and facilitates the evolution of management into a hybrid socio-intercultural, institutional, and organizational framework, as well as local hybrid economics (Altman & Cochrane, 2003; Altman, 2001).

Community mapping and social planning systems, supported by an analysis of the impact on the assessment framework of values, rights, and experiences in natural resources management, enable decisions to be made based on public participation, self-determination, and self-governance (Loomis, 2000; Ridder & Pahl-Wostl, 2005; Xanthaki, 2007).

A sustainable development framework should promote the value of alternative and divergent thinking, taking into account the pluralism of worldviews, cultural diversity, and the revival of local community environmental culture. This includes the restoration of traditional systems of natural resource management and the incorporation of these concerns into the development of environmental policies that address sustainable development issues. Cultural diversity is particularly evident in urban areas, where policies have been implemented to manage cultural pluralism as a valuable resource.

When it comes to empowering individuals through socio-intercultural partnerships and involving them in decision-making processes, systemic thinking, problem-solving, and cause-effect approaches have their limitations. Therefore, it is important to focus on capacity building and the promotion of socio-intercultural partnerships to influence different worldviews. This will help develop the necessary abilities and core competencies within the framework of socio-intercultural dialog, which in turn will inform alternative pathways towards a sustainable society.

In order to enable the necessary dialog, self-determination in governance processes offers various possibilities through negotiations between local governments and communities. Indigenous and local authorities face challenges in achieving recognition and autonomy in natural resource management systems and resource governance.

Natural resources management governance requires national, local, and community institutions to be more democratic, decentralized, and mutually supportive (Agrawal & Ribot, 1999). The design and assessment of natural resources management governance institutions can be based on governance principles that provide normative statements on the exercise of power, steering, and governance of actors in achieving their objectives. Both government and non-government institutions play crucial roles in multilevel governance processes aimed at developing natural resources management and governance, guided by the practice of good organizational governance principles. Multilevel environmental governance should ensure fairness in resource allocation and address the overlapping interests of public, private, and voluntary actors, and stakeholders.

Socio-intercultural sustainable multi-level governance structures promote the application of principles based on democratic cooperation, creating opportunities for socio-intercultural dialog and debate. A socio-intercultural sustainable multi-level governance model should consider societal diversity and promote more participatory democracy in decision-making, taking into account the territorial dimension and being inclusive of regional and local authorities.

The interdependencies among the natural resources sectors are managed within a framework of cooperation and participatory governance. This framework aims to create institutional arrangements that foster spatial and jurisdictional policies and programs, coordinating decisions and activities to overcome differences among actors. This model is more bottom-up oriented and promotes multi-level governance, rejecting top-down unilateral decisions. It provides the basis for more participatory decision-making processes and tools, fostering prospects for socio-intercultural dialog.

The multilevel natural resources management governance is an approach based on governance principles such as legitimacy, accountability, transparency, fairness, inclusiveness, capability, adaptability, and integration. These principles serve as normative guidance for organizations and local communities. The governance of natural resources management recognizes the importance of adaptability in complex and uncertain environments. It aims to develop management capacities and provides a framework of adaptive governance and flexibility to respond to contextual threats and opportunities. Organizational governance problems may arise from a lack of administrative and management capacities. In such cases, it is necessary to improve management functions, internal governance systems, and the legitimacy of all stakeholders involved in the process.

This adaptive governance approach provides the organizational structure, motivation, organizational learning, and policy design to develop instruments and indicators for monitoring and evaluating the outcomes of good governance in natural resources management. Adaptive governance is based on utilizing information for planning, policymaking, and management, which helps to navigate the uncertainty and complexity arising from various sources. It involves meta-learning and self-flexibility to change perspectives, capacities, methods, means, and organizational objectives (Schleicher-Tappeser & Strati, 2004). This approach also supports the development of instruments and indicators for monitoring and evaluating the outcomes of good governance in natural resources management.

Motivational and meta-cognitive cultural intelligence are mechanisms of social relations that have a moderate impact on the relationship between sustainable collaboration and contract governance. A high level of metacognitive cultural intelligence, supported by socio-intercultural ties, facilitates a creative approach to managing and resolving conflicts. The use of outcome performance indicators, through benchmarking, allows for tracking the governance of natural resources management and promotes transparency and accountability.

The interdependency among community members and their concerns and issues related to environmental challenges and coordination across different levels of natural resources management governance, as well as other policy and spatial domains, is evident. Nation-state governmental institutions and agencies often lack the socio-intercultural capabilities required to effectively address the demands of local communities and indigenous groups involved in natural resources management. Therefore, it is crucial for communities to actively participate in organizing for the sustainable utilization of natural resources, engaging in planning processes that involve all relevant actors and stakeholders through collective action. This collaborative approach is essential for effectively addressing sustainability concerns and issues.

Collective action has the potential to overcome non-sustainable patterns of natural resources management, leading to more sustainable governance practices (Premchander et al., 2003). Community sustainable governance is multidimensional and influenced by the systemic environmental dynamics characterized by uncertainty, complexity, tensions, and conflicting risks (Salwasser, 2004). Customary-based governance and social norms governance may encounter incidents, disputes, and tensions that can arise between different modes of governance. However, these tensions can be resolved through adaptive procedures. They may arise, for example, due to the competition for access to resources.

The deficits in socio-intercultural capacity among individuals, communities, agencies, and institutions hinder the operational effectiveness and socio-intercultural communication of natural resources management systems. To promote the development of emerging hybrid socio-intercultural systems, it is necessary to invest in enhancing the competencies and training of resources dedicated to new values, knowledge, and skills that are adapted to the socio-cultural and biophysical components of natural resources management, ensuring adaptive systems (Yunupingu & Muller, 2009) in spaces for self-management, self-determination, and accountability.

A comprehensive governance training program is required for community committees and members to share resources and achieve community goals by fostering relationships and partnerships among all involved parties.

The implementation of a socio-intercultural natural resource management governance system faces several challenges (Lunkapis, 2010). The socio-intercultural sustainable natural resources management system must analyze foundational values and consider security, autonomy, and sustainability issues within its context to determine the policies, programs, and procedures of the governance system (Cornell, 2006).

The development of principles for good governance in natural resources management and development can be supported by conducting a literature review, involving a panel of experts who can refine and test these principles (Linstone & Turoff, 1975). Governing bodies should have access to all available sources of knowledge and resources to make informed decisions and generate solutions for designing and implementing a decentralized governance system for natural resources management. This approach should embrace collaboration and avoid top-down or vertical governance structures.

Socio-intercultural sustainable natural resources management systems should be a rigorous and coherent discipline, accessible to local and global economic, social, political, and environmental processes. These systems should be responsive and accountable to the people, institutions, and groups involved at different levels of governance. The lack of socio-intercultural capacity in state governments can lead to a deficit in sustainable socio-intercultural governance, resulting in marginalization and exclusion, as measured by the impact on rights, property, and natural resources management.

2.5 DISCUSSION AND CONCLUDING REMARKS

Responsible socio-intercultural governance and management of natural resources must consider the characteristics of organizational culture and the leadership challenges involved in promoting, implementing, and sustaining socio-intercultural governance change in a dynamic world. Community leadership, management, and governance can revitalize evolving regional and local governance structures and their ongoing role in the local governance system. The socio-intercultural governance and management system of natural resources should have checks and balances to address the conflicts between the dominant and hegemonic Western worldview and the perspectives of local indigenous people.

Governance initiatives and solutions to negotiate and reconcile the differences between two governance cultures must be legitimate and effective, recognized by both the state and local indigenous communities, leading to practical outcomes. Resolving the tensions arising from complex cultural diversity requires an institutional and policy framework supported by state intervention as the most influential agent, facilitating negotiations across local governance systems to achieve mutually beneficial outcomes. Different governance cultures need to be reconciled through a process-oriented approach, aiming to achieve culturally legitimate solutions that are supported by community members and subject to discussions and negotiations.

The rationale behind local government strategies, policy processes, and goals has an impact on the legitimacy, relationships, and effectiveness of indigenous community governance. Indicators of outcome performance, as measured through benchmarking, enable tracking of natural resources management governance, and promote transparency and accountability. However, direct and unilateral interventions by the state government regarding self-governance, self-determination, and capacity building are often poorly implemented, dysfunctional, or absent, resulting in a failure to deliver socio-economic outcomes.

Dysfunctional and weak local indigenous governance undermines self-determination outcomes and is likely to create frustration within the local government itself. Poor governance operations have negative impacts on the daily experiences of organizations and communities. General recommendations for governance may be overly simplistic and not necessarily applicable to specific circumstances, given the complexity of issues and concerns related to the outcomes, making them difficult to practice. When there are expressions of discontent within the community, the internal structure of governance within organizations needs to be reformed. Additionally, governance arrangements should be reformed when the review process indicates inefficiencies in delivering services to communities.

In order to reconnect worldviews, values, responsibilities, and normative aspects of socio-intercultural interaction, natural resource management must involve a more active role of communicative action in the transfer of knowledge and mutual learning. This should be done in accordance with the visions, values, and needs of the different actors involved.

KEYWORDS

- **community leadership**
- **marginalization**
- **natural resources**
- **socio-intercultural interaction**
- **socio-intercultural natural resources management**
- **sustainability**
- **sustainable governance**
- **sustainable natural resources management**

REFERENCES

Agius, P., Davies, J., Howitt, R., & Johns, L., (2002). Negotiating comprehensive settlement of native title issues: Building a new scale of justice in South Australia. *Land, Rights, Laws: Issues of Native Title, 2*(20), 1–12.

Agrawal, A., & Ribot, J. C., (1999). Accountability in decentralization: A framework with South Asian and West African cases. *J. Dev. Areas, 33*(4), 473–502.

Allenby, B., (2006). Macroethical systems and sustainability science, *Sustainability Science, 1*(1), 7–13.

Altman, J. C., & Cochrane, M., (2003). *Sustainable Development in the Indigenous-Owned Savanna: Innovative Institutional Design for Cooperative Wildlife.*

Altman, J. C., (2001). *Sustainable Development Options on Aboriginal Land: The Hybrid Economy in the Twenty-First Century.* CAEPR Discussion Paper No. 226, CAEPR, ANU, Canberra.

Ansell, S., & Koenig, J., (2011). CyberTracker: And integral management tool used by rangers in the Djelk Protected Area, central Arnhem Land, Australia. *Ecological Management & Restoration, 12*, 13–25.

Australian Government, (2013). *Monitoring, Evaluation, Reporting and Improvement Strategy–Caring for Our Country and the Biodiversity Fund.* Department of the Environment, Commonwealth Government of Australia: Canberra. Available from URL: http://www.nrm.gov.au/system/files/resources/c2f28fc7-cb08-4951-a10a-fce03b891cf7/files/2013-18-meri-strategy.pdf (accessed on 09 March 2024).

Barbour, W., & Schlesinger, C., (2012). Who's the boss? Post-colonialism, ecological research and conservation management on Australian Indigenous lands, *Ecological Management & Restoration, 13*(1), 36–41.

Berkes, F., (2009). Evolution of co-management: Role of knowledge generation, bridging organizations and social learning. *Journal of Environmental Management, 90*(5), 1692–1702.

Bernstein, S., & Cashore, B., (2007). Can non-state global governance be legitimate? *Regulation and Governance, 1*(4), 347–371.

Bernstein, S., (2001). *The Compromise of Liberal Environmentalism.* New York: Columbia University Press.

Borrini-Feyerabend, G., Johnson, J., & Pansky, D., (2006). Governance of protected areas. In: Lockwood, M., Worboys, G. L., & Kothari, A., (eds.), *Managing Protected Areas: A Global Guide* (pp. 116–145). London: Earthscan.

Cole, D., (2011). *From Global to Polycentric Climate Governance.* Working Paper 2011/30 (Florence: European University Institute, Robert Schuman Centre for Advanced Studies Global Governance Program 2011/30, 2011), Hoffmann, Climate governance. http://ssrn.com/abstract=1858852 (accessed on 09 March 2024).

Cornell, S., (2006). *What Makes First Nations Enterprises Successful? Lessons from the Harvard Project.* Joint Occasional Papers on Native Affairs. Cambridge, and Tucson: The Harvard Project on American Indian Economic Development and the Native Nations Institute for Leadership, Management and Policy.

Costanza, R. F., Andrade, P., Antunes, M., Van, D. B. M., Boesch, D., Boersma, D., Catarino, F., et al., (1999). Ecological economics and sustainable governance of the oceans. *Ecolo. Econ., 31*, 171–187.

Dawson, T., Rounsevell, M., Kluvánková-Oravská, T., Chobotová, V., & Stirling, A., (2010). Dynamic properties of complex adaptive ecosystems: Implications for the sustainability of service provision. *Biodiversity and Conservation, 19*, 2843–2853. http://dx.doi.org/10.1007/s10531-010-9892-z.

Dovers, S., (2005). *Environment and Sustainability Policy: Creation, Implementation, Evaluation.* Sydney, Australia: Federation Press.

Ens, E. J., Finlayson, M., Preuss, K., Jackson, S., & Holcombe, S., (2012a). Australian approaches for managing 'country' using indigenous and non-indigenous knowledge. *Ecological Management & Restoration, 13*(1), 100–107.

Ens, E. J., Finlayson, M., Preuss, K., Jackson, S., & Holcombe, S., (2012b). Australian approaches for managing 'country' using Indigenous and non-Indigenous knowledge. *Ecological Management & Restoration, 13*, 100–107.

Ens, E., & McDonald, T., (2012). Caring for country: Australian natural and cultural resource management. *Ecological Management & Restoration, 13*(1), 1–1.

Esteva, G., (1987). Regenerating people's space, *Alternatives, XII*, 125–152.

Falkner, H. S., & Vogler, J., (2010). International climate policy after Copenhagen: Towards a "building blocks" approach. *Global Policy, 1*(3), 252–262.

Gerritsen, R., & Straton, A., (2007). Coping with a tragedy of the Australian aboriginal common. In: Smajgl, A., & Larson, S., (eds.), *Sustainable Resource Use: Institutional Dynamics and Economics.* Earthscan, London.

Gibbs, L. M., (2009). Water places: Cultural, social, and more than human geographies of nature, *Scottish Geographical Journal, 125*(3, 4), 361–369.

Gluckman, M., (1968). The utility of the equilibrium model in the study of social change, *American Anthropologist, 70*(2), 219–237.

Head, B., & Ryan, N., (2004). Can co-governance work? Regional natural resource management in Queensland, Australia. *Society Econ., 26*, 361–382.

Hill, R., Grant, C., George, M., Robinson, C., Jackson, S., & Abel, N., (2012). A typology of indigenous engagement in Australian environmental management: Implications for knowledge integration and social-ecological system sustainability. *Ecology and Society, 17*(1), 23. http://dx.doi.org/10.5751/ES-04587–170123.

Hill, R., Pert, P., Davies, J., Robinson, C., Walsh, F., & Falco-Mammone, F., (2013). *Indigenous Land Management in Australia: Extent, Scope, Diversity, Barriers, and Success Factors.* CSIRO Ecosystem Sciences, Cairns.

Hoffmann, B. D., Roeger, S., Wise, P., et al., (2012). Achieving highly successful multiple agency collaborations in a cross-cultural environment: Experiences and lessons from Dhimurru Aboriginal Corporation and partners. *Eco-logical Management & Restoration, 13*, 42–50.

Howitt, R., & Lunkapis, G. J., (2010). Coexistence: Planning and the challenge of indigenous rights. In: Hillier, J., &. Healey, P., (eds.), *Ashgate Research Companion to Planning Theory* (pp. 109–133). Farnham: Ashgate.

Howitt, R., & Suchet-Pearson, S., (2003). Ontological pluralism in contested cultural landscapes. In: Anderson, K., Domosh, M., Pile, S., & Thrift, N., (eds.), *Handbook of Cultural Geography* (pp. 557–569). London: Sage.

Howitt, R., & Suchet-Pearson, S., (2006). Rethinking the building blocks: Ontological pluralism and the idea of 'management. *Geografiska Annaler: Ser. B, Human Geography, 88*(3), 323–335.

Howitt, R., (2012). Sustainable indigenous futures in remote indigenous areas: Relationships, processes, and failed state approaches, *Geo Journal, 77*(6), 817–828.

Howitt, R., Doohan, K., Suchet-Pearson, S., Cross, S., Lawrence, R., Lunkapis, G. J., Muller, S., et al., (2013). socio-intercultural capacity deficits: Con-tested geographies of coexistence in natural resource management. *Asia Pacific Viewpoint, 54*, 126–140.

Hughes, H., (2005). *Policies Entrench Poverty*. The Australian. Available at http://www.cis.org.au/executive_highlights/EH2005/eh30205.html (accessed on 09 March 2024).

Jackson, M. V., Kennett, R., Bayliss, P., et al., (2015). Developing collaborative marine turtle monitoring in the Kimberley region of Northern Australia. *Ecological Management & Restoration, 16*, 163–176.

Lane, M. B., (2003). Decentralization or privatization of environmental governance? Case analysis of bioregional assessment in Australia. *J. Rural Stud., 19*, 283–294.

Lenton, R., (2002). Managing natural resources in the light of climate variability. *Nat. Resources Forum, 26*, 185–194.

Linstone, H. A., & Turoff, M., (1975). *The Delphi Method: Techniques and Applications*. Reading, MA: Addison-Wesley.

Loomis, T. M., (2000). Indigenous populations and sustainable development: Building on indigenous approaches to holistic, self-determined development. *World Development, 28*(5), 893–910.

Lunkapis, G. J., (2010). *Understanding Land Use and Natural Resources (LUNaR) Governance: A Case Study of Sabah, Malaysia*. Sydney: Department of Environment & Geography: Macquarie University.

McKenna, M., (2004). *The Need for a Reconciled Republic*. Manning Clark House Inc, Canberra, Available at: http://www.manningclark.org.au/papers/reconciled_republic.htm (accessed on 09 March 2024).

Meridian Institute, (2001). *Comparative Analysis of the Forest Stewardship Council and the Sustainable Forestry Initiative Certification Programs*. Washington, DC: Meridian Institute.

Morrison, T. H., McDonald, G. T., & Lane, M. B., (2004). Integrating natural resource management for better environmental outcomes. *Austr. Geogr., 35*, 243–258.

Muller, S., (2008c). *Making Space to Care for Country*. Department of Human Geography. Sydney: Macquarie University.

Muller, S., (2012). Two ways: Bringing indigenous and non-indigenous knowledges together. In: Weir, J., (ed.), *Country, Native Title, and Ecology* (pp. 59–79). Canberra: ANU ePress.

Muller, S., (2014). Co-motion: Making space to care for country. *Geoforum, 54*, 132–141.

Natcher, D. C., Davis, S., & Hickey, C. G., (2005). Co-management: Managing relationships, not resources. *Human Organization, 64*(3), 240–250.

Orts, E., (2011). *Climate Contracts*. University of Pennsylvania Institute of Law, and Economics research paper, 11–31. http://papers.ssrn.com/sol3/papers.cfm?abstract_id=1814504 (accessed on 09 March 2024).

Ostrom, E., (1990). *Governing the Commons: The Evolution of Institutions for Collective Action*. Cambridge, UK: Cambridge University Press.

Premchander, S., Jeyaseelan, L., & Chidambranathan, M., (2003). In search of water in Karnataka India—Degradation of natural resources and the livelihood crisis in Koppal *District. Mount. Res. Dev., 23*(1), 19–23.

Reinicke, Wolfgang, H., Francis, D., Thorsten, B., Jan, M. W., Beth, W., & John, G., (2000). *Critical Choices: The United Nations, Networks, and the Future of Global Governance*. Ottawa: International Development Research Centre.

Ridder, D., & Pahl-Wostl, C., (2005). Participatory integrated assessment in local level planning, *Regional Environmental Change, 5*(4), 188–196.

Salwasser, H., (2004). *Confronting the Implications of Wicked Problems: Changes Needed in Sierra Nevada National Forest Planning and Problem Solving*. USDA Forest Service General Technical Report PSW-GTR-193. Albany, NY: USDA Forest Service.

Sanders, W., (2006). *Being a Good Senior Manager in Indigenous Community Governance: Working with Public Purpose and Private Benefit.* CAEPR Discussion Paper No. 280, CAEPR, ANU, Canberra.

Schäferhoff, M., Campe, S., & Kaan, C., (2009). Transnational public–private partnerships in international relations: Making sense of concepts, research frameworks, and results, *International Studies Review, 11*(3), 451–474.

Schleicher-Tappeser, R., & Strati, F., (2004). Structural funds and sustainable development: The SQM approach. *Innovation, 17*, 75–94.

Sinclair, A., (2007). *Leadership for the Disillusioned: Moving Beyond Myths and Heroes to Leading That Liberates*, Allen & Unwin, Crows Nest.

Smith, D. E., (2007). Networked governance: Issues of process, policy, and power in a West Arnhem Land regional initiative. *Ngiya: Talk the Law, Governance in Indigenous Communities, 1*, 24–52.

Sobeck, J., & Agius, E., (2007). Organizational capacity building: Addressing a research and practice gap, *Evaluation and Program Planning, 30*(3), 237–246.

Suchet, S., (1999). *Situated Engagement: A Critique of Wildlife Management and Postcolonial Discourse*. Sydney: Department of Human Geography, Macquarie University.

Suchet-Pearson, S., & Howitt, R., (2006). On teaching and learning resource and environmental management: Reframing capacity building in multicultural settings. *Australian Geographer, 37*(1), 117–128.

Turnhout, E., (2010). Heads in the clouds: Knowledge democracy as a Utopian dream. In In 't Veld, R. J., (ed.), *Knowledge Democracy* (pp. pp. 25–36). Heidelberg: Springer Berlin.

Weir, J., (2009). *Murray River Country: An Ecological Dialogue with Traditional Owners*. Canberra: Aboriginal Studies Press.

Weir, J., (2011). Water planning and dispossession. In: Connell, D., & Grafton, R. Q., (eds.), *Basin Futures* (pp. 179–191). Canberra: ANU ePress.

Witte, J. M., Charlotte, S., & Thorsten, B., (2003). *Progress or Peril? Partnerships in Global Environmental Governance*. Washington, DC, and Berlin: Global Public Policy Institute.

Wohling, M., (2009). The problem of scale in indigenous knowledge: A perspective from Northern Australia, *Ecology and Society, 14*(1), 1. Retrieved from: http://www.ecologyandsociety.org/vol14/iss11/art11 (accessed on 09 March 2024).

Wright, S., Lloyd, K., Suchet-Pearson, S., Burarrwanga, L., & Tofa, M., (2012). Telling stories in, through and with Country: Engaging with indigenous and more-than human methodologies at Bawaka, NE Australia, *Journal of Cultural Geography, 29*(1), 39–60.

Xanthaki, A., (2007). *Indigenous Rights and United Nations Standards Self-determination, Culture and Land.* Cambridge, UK. New York: Cambridge University Press.

Yunupingu, D., & Muller, S., (2009). Cross-cultural challenges for indigenous sea country management in Australia. *Australasian Journal of Environmental Management, 16*(3), 158–167.

CHAPTER 3

Internationalization of Startups Through Entrepreneurial Marketing and Temporary Cessation of Overseas Sales

FAKHRADDIN MAROOFI[1] and SEYED FARHAD HOSSEINI[2]

[1]*Department of Management, Faculty of Humanities and Social Sciences, University of Kurdistan, Sanandaj, Iran*

[2]*Department of Business Management and Accounting, Faculty of Management and Accounting, Razi University, Kermanshah, Iran*

ABSTRACT

This research utilizes the effectiveness perspective to enhance understanding of the entrepreneurial marketing practices of companies that undergo rapid internationalization followed by temporary cessation of international sales. The significance of this current research lies in its departure from previous studies that primarily focused on the continuation of operations in foreign markets, rather than the decision to exit the international market. Prior to international planning, semi-structured interviews were conducted with managers of 12 small knowledge-based startups in Iran. All shareholders demonstrated active involvement, risk-taking, innovation exploitation, and opportunistic behavior, yet their resource utilization fell short in terms of creating sufficient "client value." The findings of this research indicate that certain aspects of an effective approach surpass others in specific areas, and the concept of "cost-effectiveness" encompasses both financial and non-financial considerations. This study offers unique insights into the necessity for corporate decision-makers to swiftly internationalize, validate new client discovery strategies, and acknowledge the potential loss of specific clients,

Contemporary Digital Transformation and Organizational Effectiveness in Business 4.0.
Nilanjan Ray, Sudin Bag, & Tapas Kumar Chatterjee (Eds.)

even if only temporarily, in order to facilitate sustainable entrepreneurial marketing activities.

3.1 INTRODUCTION

3.1.1 PURPOSE OF THE RESEARCH

Entrepreneurial marketing (EM) is a process of orientation aimed at capitalizing on growing opportunities and investments through innovation, creativity, sales, market penetration, networking, and flexibility. By fostering relationships, EM creates value for clients (Hills & Hultman, 2011). It represents a departure from traditional marketing perspectives, as it emerges from the performance of companies operating in uncertain environments. However, many startups struggle in such uncertain contexts. Therefore, understanding the "context" becomes crucial in analyzing company behavior, as unique insights from specific contexts can be instrumental in gaining knowledge. This study focuses on international market exit as an attractive opportunity to investigate entrepreneurial marketing behavior. Preliminary research has explored the dimensions of entrepreneurship and market orientation, including service provision to foreign markets and international market withdrawal (Yayla, Yeniyurt, Uslay, & Cavusgil, 2018). Previous research has discussed the concepts of "uncertainty" and "risk" (McKelvie, Haynie, & Gustavsson, 2011). In this study, the term "risk" refers to the perception of decision makers. Specifically, "uncertainty" in the initial study implies that decision makers cannot assign specific outcomes to different results, whereas "risk" suggests that decision makers can act based on certain probabilities, informed by research and experience. Considering the potential risks, Johanson & Vahlne (1977) examined internationalization based on increased knowledge and commitment (Johanson & Vahlne, 1977).

In contrast, some companies are rapidly becoming international and require factors such as achieving operational status and accessing stakeholders such as startups and reward institutions (Crick & Crick, 2018; Hewerdine, Rumyantseva, & Welch, 2014). Previous studies have used terms like "New International Investments" to categorize companies that have rapidly expanded internationally, so none of the agreed criteria is unique to this category. Current research uses the term new international investments to rank the companies in question based on their rapid internationalization. However, their export-to-sales ratio (international scale) was small. Additionally, the scale (or geographic coverage) of foreign markets prior to their entry into international markets was also limited. As a result, it will be challenging for them to commit to overseas sales and expand their market reach. Limited research exists on the ongoing behavior of new international investments in the context of international activities (Ibeh, Jones, & Kuivalainen, 2018). There is a gap in understanding the performance of new international investments that rapidly withdraw from foreign markets instead of sustaining international operations. This study aims to bridge this gap by examining the aspects of new international entrepreneurial marketing investments that facilitate sustainable operations after exiting foreign markets and before planned re-internationalization. In other words, it focuses on entrepreneurial marketing decisions during the "interval" between exiting a foreign market and re-entering an international market, which is crucial for strategy re-evaluation (Welch & Welch, 2009). Research indicates that internationalization is not solely a forward-looking process (Bernini, Du, & Love, 2016; Yayla, Yeniyurt, Uslay, & Cavusgil, 2018). However, limited perspectives exist on this matter, highlighting the need for further research on new international investments that rapidly exit the international market (Yayla, Yeniyurt, Uslay, & Cavusgil, 2018). On the other hand, it is not unlikely for startups to enter these markets hastily only to withdraw quickly. Particularly, startups prioritize minimizing potential losses over potential profits when targeting end consumers (Crick & Crick, 2018b). These considerations lead to the following research questions.

- **Research Question:** What role does the entrepreneurial marketing decision play for shareholders in emerging international investments that are rapidly exiting foreign markets?

3.1.2 PARTICIPATION

This research builds upon the seven dimensions of entrepreneurial marketing behavior mentioned by Morris et al. (2002), in line with existing knowledge about sustainable entrepreneurial marketing behavior of new international investments (Morris, Schindehutte, & Laforge, 2002). These dimensions include an active orientation, opportunity-based behaviors, client size, innovation-based behaviors, risk management, resource use, and value creation behaviors. Utilizing the effectiveness perspective as a scientific foundation, we extensively utilized Sarasvathy's original research (2001) and discovered aspects of the five-dimensional approach to effectiveness that are apparent in entrepreneurial decision-making (Prashantham, Kumar, Bhagavatula, & Sarasvathy, 2018). Our initial involvement, which we identify in specific circumstances, can be a significant factor in the effectiveness perspective of decision making. In a particular area of this research, entrepreneurs focused on the "cost-effective principle" when deciding to exit the international market. Other decision-making elements become evident in the initial internationalization and subsequent considerations of re-internationalization. The second intervention involves identifying the various dimensions of the concept of "cost-effective loss" perceived by entrepreneurs. Thirdly, we participate in the process of discovering clients in the new product market by demonstrating how the various dimensions of entrepreneurial marketing need to come together to help sustain entrepreneurial marketing behavior. Considering the risk/reward consideration, we have shown that all shareholders are proactive, utilizing the accepted risk of innovation, and intervening with opportunistic behavior and resource use (entrepreneurial dimensions centered on entrepreneurial marketing behavior). However, a decision that addresses "all" aspects of entrepreneurial marketing is also evident. This entrepreneurial marketing behavior has led to a temporary exit from the foreign market and the need to evaluate perceived risk/reward strategies and cost-effective losses. In summary, this research provides unique insights into the need for decision-makers in companies that have rapidly internationalized to validate new client discovery strategies. Additionally, why is it potentially beneficial for shareholders, especially in fast-growing international companies, to temporarily lose some of their international clients, thereby facilitating their sustainable entrepreneurial marketing activities? This is somewhat at odds with some existing research.

3.2 LITERATURE REVIEW

3.2.1 SCOPE OF ENTREPRENEURIAL MARKETING

The past few decades have witnessed the emergence of knowledge that has widely regarded entrepreneurial marketing as a distinct area of research in the literature on entrepreneurship and marketing (Hills, Hultman, & Miles, 2008; Miles, Gilmore, Harrigan, Lewis, & Sethna, 2014). Rapid technological changes and the limited impact of strategies and techniques have resulted in uncertainty. Despite several researchers emphasizing the need for further development of the entrepreneurial marketing framework (Ionita, 2012), the literature in this field remains fragmented. Bjerke and

Hultman highlighted the necessity for theory-based research in the field of entrepreneurial marketing (Bjerke & Hultman, 2002). Ionita (2012) identifies this gap in a way that, despite attracting researchers from various disciplines such as marketing, entrepreneurship, economics, psychology, and sociology, there is still room for work and development in this field (Ionita, 2012). A literature review reveals that, despite numerous studies on the identification and prioritization of entrepreneurial marketing components, the relationship between these components and the hierarchy of these relationships remains unclear. These circumstances, coupled with the lack of comprehensive research in the field of entrepreneurial marketing in the country, have prompted researchers to develop a comprehensive model of entrepreneurial marketing to fill this gap and contribute to the advancement of the literature on this concept. Therefore, entrepreneurial marketing considers perspectives related to entrepreneurship and marketing. However, researchers point out that the perspectives within the field of entrepreneurial marketing are diverse and in practice, lead to differences in findings related to the implementation of entrepreneurial marketing (Hansen & Eggers, 2010). In other words, existing research examines entrepreneurial marketing from various perspectives, such as interdisciplinary perspectives or perspectives on entrepreneurship and marketing. There are multiple definitions of entrepreneurial recovery (Hills & Hultman, 2011). In this study, we adopt the definition provided by Hills & Holtman (2011), which focuses on the investment growth and the implementation of entrepreneurial marketing activities in larger companies (Miles & Darroch, 2006). Khayatian et al. (2015), through theoretical foundations and literature review, demonstrate that the concept of knowledge-based companies overlaps with the concepts of small and medium-sized companies and start-ups (Khayatian, Tabatabayan, Amiri, & Eliassy, 2015). Small and medium-sized companies, in general, and knowledge-based companies, in particular, play a crucial role in economic growth and development by commercializing their knowledge achievements (Khayatian, Tabatabayan, Amiri, & Eliassy, 2015; Roundy, 2017).

Becherer et al. (2012) examined the impact of entrepreneurial marketing on the quantitative and qualitative outputs of small and medium-sized companies (Becherer, Helms, & McDonald, 2012). This study is based on a quantitative method and is of the survey type. Factor analysis and regression techniques were used to analyze the data. The results of this study indicate that entrepreneurial marketing has an effect on company outputs, particularly value creation, as entrepreneurial marketing encompasses behavior that is not limited to businesses with commercial goals but can also be applied to organizations with "social" goals (Becherer, Helms, & McDonald, 2012; Roundy, 2017). However, this research includes start-ups with commercial goals. The concept of entrepreneurial marketing has been applied in various contexts. Some view entrepreneurial marketing for small and medium-sized businesses as an enhanced process (Morrish, 2011). Others consider it suitable for emerging firms, while some apply it to large technology-based companies (Azar, Sadeghi, & Kordnaeij, 2012). Recent research also suggests that entrepreneurial marketing is not necessarily dependent on the age and size of the company (Schmid, 2012) and represents an opportunistic approach where marketers pursue innovation. In the context of this research, environmental challenges associated with early internationalization present potential obstacles to realizing this value unless there is sufficient support (Zhu & Matsuno, 2016; Ries, 2011). Gassmann et al. (2014) found that the majority of companies employ entrepreneurial marketing strategies in dynamic environments (Gassmann, Frankenberger, & Csik, 2014). This discovery aligns with recent research, confirming the potential for internationalization growth in evolving business model environments as a recurring process undertaken by decision-makers to generate excitement in the product market (Morrish, 2011). Marketing influences entrepreneurship by utilizing tools, concepts, and theories to support the growth of existing businesses and the creation of new employment opportunities. Ahmadi & O'Cass (2016) focused on entrepreneurial marketing activities by integrating entrepreneurial and market orientation to extract innovation through embracing new technology investments, identifying latent needs, and developing marketing programs in novel business environments (Ahmadi & O'Cass, 2016). Although marketing and entrepreneurship are both change-oriented, opportunistic, and possess an innovative approach to management (Collinson & Shaw, 2001), they have a positive impact on exploration and exploitation activities, thereby enhancing performance and the potential role of market capabilities. While market orientation and entrepreneurship may appear as distinct concepts, their ability to interact and combine is crucial, as this interaction significantly influences business performance (Morrish, 2011). Matsuno et al. (2002) argue that these two approaches not only affect organizational performance but also each other (Matsuno, Mentzer, & Ozsomer, 2002). However, there are entrepreneurial marketing initiatives aimed at building capabilities and leading to diverse strategies for improving performance among decision makers (Bell, Crick, & Young, 2004; Bell, McNaughton, Young, & Crick, 2003). Additionally, Mirvahedi & Morrish (2017) discussed the concept of "entrepreneurial luck" and found that entrepreneurs exhibit behavior in searching for unspecified opportunities through regular or random means, which may occur at varying levels of knowledge (Mirvahedi, 2017). The occurrence is specified. In summary, understanding entrepreneurial marketing decisions provides valuable insights for formulating strategies in specific areas. Hills & Hultman (2011) summarized in their research that the development of the theory of effectiveness (Sarasvathy, 2001) helps researchers make decisions based on risk/uncertainty conditions and consequently, understand the important dimensions of entrepreneurial marketing behavior. Therefore, a suitable perspective in this research seems.

3.2.2 EFFECTIVENESS-BASED DECISION MAKING

The point of view in this study includes decisions based on "effectiveness" (Sarasvathy, 2001). Entrepreneurs prioritize minimizing potential losses over maximizing profits, demonstrating their focus on cost-effectiveness. Entrepreneurs collaborate with their partners to jointly shape future strategies. Are entrepreneurs open-minded enough to be astonished by the potential opportunities that arise? According to Sarasvathy (2001), there is a discussion on causal competencies versus effectiveness approaches, where causal decision-making involves considering different options and working towards a goal. Furthermore, effective decision-making can be complemented in various ways, depending on the available options (Arend, Sarooghi, & Burkemper, 2015; Read, Sarasvathy, Dew, & Wiltbank, 2016). Effective causal decision-making proves valuable in unpredictable future scenarios. Previous research on international companies supports the effectiveness approach (Andersson, 2011; Galkina & Chetty, 2015; Gabrielsson & Gabrielsson, 2013). For instance, Andersson (2011) explores the knowledge and expertise required by entrepreneurs in international companies. This is particularly relevant for startups that rely on the skills of their small management teams. Crick & Crick (2018) argue that resource-based startups can enhance their network by engaging with new partners such as investors, resources, and unpredictable capabilities (Crick & Crick, 2018b) (Table 3.1).

TABLE 3.1 Details of Selected Companies

Company	Internationalization Path			
	TI to Inter	**Pro**	**Expo**	**Viewpoint**
No. 1	Year 3	Markazi province	EXP	The export of Markazi province is superficial research and cannot have added value because this technology was not needed after the first initial order.
No. 2	Year 3	Tehran province	EXP	Tehran's exports included petrochemical equipment, but although the technology was good, it was cheap, and the possibility of repeating orders was low due to sanctions.
No. 3	Year 1	Southern province	EXP	Exports from the southern province, including light bulbs, were cheap and had many problems due to sanctions.
No. 4	Year 3	Khorasan Razavi	EXP	Khorasan Razavi exports, and this technology had no added value because the competitor offered a modified and cheaper version.
No. 5	Year 2	Zanjan province	EXP	Zanjan province exports included detergent powder, but customers rejected the powder due to lack of standard.
No. 6	Year 2	Tehran province	EXP	Tehran province exports included new technology and were rejected by the importer due to sanctions and complexity.
No. 7	Year 1	Zanjan province	EXP	Zanjan province exports included food, but customers rejected it because of its poor quality.
No. 8	Year 1	Tabriz province	EXP	The export of Tabriz province included construction and decorative tiles, it is good and cheap and has added value.
No. 9	Year 3	Tabriz province	EXP	Tabriz province exports included technology, but customers found this technology very complex.
No. 10	Year 1	Tehran province	EXP	Tehran province exports included petro-chemical equipment, but despite the good technology, it was cheap and there were many orders.
No. 11	Year 2	Kermanshah province	EXP	Kermanshah province exports included petrochemical equipment, but despite good technology, it was cheap and had good orders.
No. 12	Year 1	Isfahan province	EXP	Isfahan province exports included petro-chemical equipment and steel, but despite good technology, it was cheap and had good orders.

3.2.3 EXITING THE INTERNATIONAL MARKET AND POTENTIAL RE-INTERNATIONALIZATION

Research primarily focuses on internationalization as an incremental process (Jones, Coviello, & Tang, 2011), which is influenced by learning and increased commitment over time (Jones & Coviello, 2005). Conversely, exiting the international market based on decision makers' perceptions may involve reducing commitment in various dimensions. Turner (2012), based on the theory of joint evaluation, identified different paths of constriction and expansion followed by selective companies. These routes are unpredictable, resulting in various issues and problems affecting relevant stakeholders. Such an approach aligns with effectiveness-based decision making. Jones et al. (2011) argue that exiting the international market is influenced by past international operations and commitments, as well as recent developments inside and outside the company (Jones, Coviello, & Tang, 2011). However, the nature of "past" activities may vary between firms and startups involved in recent research, which have rapidly internationalized with limited experience among shareholders. Thus, according to the research of Hansen & Eggers (2010) on entrepreneurial marketing behavior, attempts to understand credibility are useful for identifying market opportunities and product excitement (Hansen & Eggers, 2010). Welch & Wiedersheim-Paul (1980) state that exiting the international market is not necessarily indicative of a company's failure, but rather depends on the company's position and the goals of decision makers. Or companies may formulate their strategy based on the changing situation or other factors such as financing. According to Welch & Welch (2009, p. 567), the re-internationalization of companies depends on various time periods, such as changes in management, ownership, or power. Therefore, the impact on re-entry depends on the re-entry process (Welch & Welch, 2009). As mentioned earlier, the conditions for entrepreneurial marketing decision-making in the recent study occur when new international investments are temporarily withdrawn from foreign markets, rather than during re-internationalization.

3.3 METHODOLOGY

3.3.1 OVERVIEW

The qualitative research method involves the collection and analysis of non-numerical data to identify different concepts, perspectives, and experiences. Qualitative research relies on high-quality sources to gather information and data. In previous research (Gilmore, 2010), the "how/why" issues have been addressed. Additionally, as demonstrated by Woodside et al. (2005), asking employees for a comprehensive explanation of what happened and why can potentially help them reflect and reveal the true and accurate points or mistakes that occurred. This study focuses on semi-structured interviews with the managers of 12 new international investments that have been temporarily withdrawn from the international market, and its nature is entirely interpretive. The semi-structured interviews, conducted in Persian, lasted for one hour. These interviews allow for exploration of the "how and why" issues, enabling a discussion of important matters that provide more information. To ensure mutual understanding between the interviewer and interviewee, common meanings were established using employee terminology. The collection of company documents and data sources was limited due to confidentiality. The data collected in these interviews includes decisions related to internationalization, management teams, exit strategies from the international market, and strategic plans for the future of the companies.

3.3.2 DATA COLLECTION

This research is limited due to the lack of access to multi-source data, so it is sufficient to conduct interviews with the managers and owners of the company (Welch, Piekkari, Plakoyiannaki, & Paavilainen-Mantymaki, 2011). Achieving the right number of cases (interviews) can be challenging in order to strike a balance between theoretical saturation and case comparison (Beverland & Lindgreen, 2010). In this study, an unconventional targeted sampling strategy was employed (Saunders, Lewis, & Thornhill, 2009). Data collection involved engaging 12 managers of new international investments who had recently exited international markets, using a snowball process. Specifically, this process relied on working networks associated with an institution where one research team operated, as well as another network, such as a business facility. Conducting interviews with key informants, despite the fact that the data originated from a single source, proved to be a valuable aspect of this study, as the perspectives of other interviewees would have added limited value to entrepreneurial marketing decisions. Furthermore, the interviewees' responses were cross-checked against financial market reviews, thereby identifying the potential for review bias. Due to the relative novelty of the companies, access to secondary data was limited. Start-ups, being relatively new and highly competitive in the market alongside large, established companies, swiftly enter or exit the international market by evaluating their strategies, often resulting in a high failure rate (Blank, 2013). New international investments from intensive knowledge sectors, such as those often referred to as "advanced technology," have the potential for rapid internationalization (Bell, Crick, & Young, 2004; Crick & Crick, 2014, 2018). According to a previous study, "rapid internationalization" requires new international investment to enter an emerging market from a fledgling sector (Jones, Coviello, & Tang, 2011). The ability to initiate the process of business internationalization relies on the presence of sustainable competitive advantage in the global market, driven by entrepreneurial institutions and embedded within the core of the business. This is also influenced by market conditions, internal competition, firm size, and existing experience. Typically, new international investments are classified as small and medium-sized businesses with fewer than 40 employees (Storey, 1994).

3.3.3 ANALYSIS

Presents qualitative research, demonstrating a level of predictability and deductive reasoning that must be followed to identify the real trend. "Problems" in qualitative research often exhibit repetition, and interviews involve inquiries from decision-makers regarding literature review matters (Timmermans & Tavory, 2012). This research on the nature of "how and why" has conducted further investigations on various topics. Consequently, this study adopted an "inductive" approach, which involved navigating between primary and secondary data to generate findings as an emerging pattern. In this study, due to the iterative process, emphasis was placed on progression. Therefore, the issue of expertise was explored, along with risk-taking, which represents one of the dimensions of entrepreneurial orientation. As a result, researchers considered factors such as circumstances, narratives, and individual involvement in the research.

Since concepts form the fundamental basis of theory, it is imperative to establish a mechanism within data theory to identify and expand upon these concepts based on their characteristics and dimensions. This mechanism is carried out through open coding, where the researcher extracts initial categories from the raw data that are relevant to the phenomenon being studied. This is achieved by posing questions about the data, comparing different cases, events, and other phenomena. Furthermore, Sinkovics &

Alfoldi (2012) emphasized the potential benefits of utilizing software in data analytics (Sinkovics & Alfoldi, 2012).

The active participation of businesses in the global market has necessitated the development of domestic technology in order to withstand intense global competition. While it may be possible to operate with existing technologies in the domestic market, the key to survival in international markets lies in the development and enhancement of advanced technologies. Therefore, in the current landscape of knowledge-based businesses, engaging in international activities is essential for elevating technological capabilities and achieving global recognition. Without it, establishing a presence at the international level is not feasible for any firm.

However, in this particular study, the authors encountered limitations due to economic sanctions, resulting in a small dataset that required manual coding. Specifically, data analysis using manual coding was integrated into three stages: open coding, axial coding, and selective coding, which aligns with the JVIA method. It is important to note that the second-order themes derived from this process were not solely based on empirical data following an inferential approach, but rather emerged from an "inductive" research design that incorporated a literature review. Second-order themes are derived from Morris et al. (2002), followed by a set of research questions (Morris, Schindehutte, & Laforge, 2002). The data's "trustworthiness" (Lincoln & Guba, 2000; Morrow, 2005), including validation, transferability, and reliability, was also taken into account. These considerations partly determine the extent to which the study's findings can be applied to other situations. For instance, to enhance the credibility of this research, a previously defined "boundary" was established, points were shared with the interviewees, and discussions were held among the researchers to address relevant issues.

3.4 FINDINGS

3.4.1 "ISSUES" DESCRIBING EXIT FROM THE INTERNATIONAL MARKET

Every production in the economy, regardless of the level of production return, necessitates prior investment. Furthermore, every investment, from its inception, requires capital security until the years when it begins to yield returns. Additionally, the lack of preparedness for internationalization impacts decision-makers in their temporary withdrawal from the international market. One of the primary causes of entrepreneurial marketing issues is the lack of credibility, including the failure to identify customer needs and provide value-added services. In fact, many businesses have rapidly expanded their international presence. In light of these circumstances, all interviewees acknowledged that even a temporary loss of a client proved beneficial in evaluating entrepreneurial marketing strategies and minimizing cost-effective losses. However, the perception of the nature of this withdrawal varied among the interviewees, although they agreed that the duration should be brief (ranging from two months to several months). Shareholders made decisions before reconsidering internationalization, and then resumed their efforts. All interviewees shared a broad objective of internationalization (somewhat aligned with causal decision-making). However, the choice of country depended on the risk/reward considerations and emerging opportunities, while remaining undecided (according to effective decision-making). One of the main issues among all interviewees was the fast pre-budget period, i.e., financially cost-effective losses. Exporting as a strategy makes it relatively easy for all companies to enter the market and then decide to exit compared to other subsidiaries. In marketing evaluations, entrepreneurs also found that the company must have the resources, capabilities, experience, and knowledge of the previous job to invest. All interviewees stated that foreign sales are a good reference for potential customers and investors. However, the first move, gaining a foreign market, increases credit in small global markets. Additionally, entering the foreign market influences the company's decisions. For example, the interviewee at Company 1 stated, "Chance plays a small role that most people (customers) interact with, so companies need to be fully prepared. However, success can be a great indicator of expansion." Therefore, the main issue among all interviewees is about cost-effective non-financial losses, especially from a credit perspective in certain small markets, in addition to potential financial losses.

However, all interviewees emphasized the importance of opportunities, stating that companies, given the characteristics of entrepreneurial marketing (Blank, 2013; Gassmann, Frankenberger, & Csik, 2014), should proactively seek to innovate, utilize their capabilities, and take calculated risks in order to develop an effective strategy or proposal. As a result of different perspectives, decisions regarding risk and reward will inevitably lead to the acceptance of unexpected orders. For instance, interviewees from Company 5 mentioned, "We take our chances through networking and open collaboration. However, it must be noted that we may not always be fortunate with our choices!" Therefore, all decision-makers who decided to exit foreign markets envisioned minimizing financial losses for stakeholders or providing additional credit to offset such losses. Furthermore, all entrepreneurs and managers initially explored various aspects of entrepreneurial marketing behavior without conducting thorough research or obtaining proper accreditation. For instance, opening an initial account was suggested as a means of utilizing personal savings, which required limited findings to reflect customer characteristics.

As stated by an interviewee from Company 10, "We have knowledge of the target customers, but we are uncertain about their exact desires and value proposition." The potential financial losses were expected to be significant for each company in terms of monetary effectiveness, as most stakeholders had collateral at stake, such as their homes, in the event of a business failure. "I had to put my house on the line, but it's a risky move," said an interviewee from Company No. 3. Considering that stakeholders were heavily invested but unable to allocate minimum funds for market research, there was a high level of accepted risk due to various environmental pressures. "Research is undoubtedly helpful, but it is costly and time-consuming, so we had to act swiftly," stated the interviewer from Company No. 7, the second start-up company. Taking into account the details of the shareholders, internationalization presents a credible opportunity in terms of credibility and significance in the target market. However, the level of importance attributed to it varies among the interviewees. In particular, the term "shareholders" refers to those individuals with whom the interviewees may wish to engage in the future, or even potential investors. For instance, the interviewee from Company 12 emphasized that expanding presence in the international market can open up new possibilities for brand introduction and collaboration. We, as active shareholders, can take advantage of this opportunity and make an investment. This is an example of the entrepreneurial marketing characteristics sought in all interviewees who have international experience. They are actively seeking additional resources, capabilities, sales, creativity, energy, experience, and loyalty to the company. They are also proactive, risk-averse, opportunistic, and resourceful. However, without proper research, credibility, and the decision of the creative team to internationalize, entering foreign markets would be a mistake considering the dimensions of entrepreneurial marketing. "We felt somewhat embarrassed," stated the interviewee from Company 12. Therefore, it is necessary to withdraw from foreign markets in order to minimize the company's losses and potential network partner customers.

3.4.2 DECISION-MAKING AND THE POTENTIAL FOR FUTURE RE-INTERNATIONALIZATION

Previous Negative experiences, following initial preparations for internationalization and the realization of "cost-effective" financial and non-financial losses, necessitate the decision to temporarily withdraw or take a break from the international market. The main objective of this study is to explore the re-internationalization of companies and strategies for restoring credibility with network partners in the future. At the early stages of starting a startup, none of the shareholders or creative teams offered to invest. The role of stakeholders was re-evaluated during the break. Ultimately, it was determined that the company would require an investor to overcome the crisis and maintain credibility after exiting the international market. One interviewee from Company 6 acknowledged, "The right person can bring more than just financial gains, such as networks and shared work experiences with others like us. They can open doors." All interviewees recognized the importance of decision-making control when investors acquire capital stock, except for potential facilitative opportunities. However, this risk/reward decision could potentially support entrepreneurial marketing activities considering the cost-effective losses. "An investor wants to make money, so we need to demonstrate market potential and showcase our exceptional team," stated an interviewer from Company No. 10. The interviewees' perception of building/redeeming credit includes the potential for future public stock offerings. "An investor wants to make money, so there may come a time when we have to sell the business or take the company public if it makes sense," mentioned an interviewee from Company 11.

The literature on entrepreneurial marketing emphasizes the importance of entrepreneurship, as companies strive to expand internationally. Therefore, it is crucial to transform goals in order to effectively guide entrepreneurial marketing activities. However, all interviewees also recognized that even if the companies operate without investor support, special measures are necessary to advance their business, including the need for international restructuring. In summary, this affects the decision-making process by learning from errors (such as non-validation), addressing revenue and credit issues, and striving to restore a positive image among shareholders. "Being successful abroad allows us to showcase our products, so we need to go back (abroad) for clients to become aware of us," said an interviewee from Company 7. The respondent acknowledged that "we have to do our homework (validation) and consider some damage control (credit recovery)."

3.5 DISCUSSION

The present study is timely and important for studying the phenomenon of internationalization because it involves entrepreneurial marketing decisions in new international investments, as well as small and medium-sized knowledge-based businesses that have rapidly exited the international market and have recently been contacted for further research (Yayla, Yeniyurt, Uslay, & Cavusgil, 2018). Previous research has primarily focused on companies that continue to operate overseas. The research question is: What role does entrepreneurial marketing decision play for shareholders in new international investments that are rapidly exiting international markets? These findings have been obtained in previous research using the perspective of effectiveness and the process of internationalization of small and medium-sized businesses, the existence of a sustainable competitive advantage in the world class (Andersson, 2011; Prashantham, Kumar, Bhagavatula, & Sarasvathy, 2018). Although existing research has shown that the focus of recent research has not been considered, it provides the basis for the ability to initiate propulsion within the entrepreneurial entity (such as international experience and managerial knowledge) and the heart of the business (such as the need to grow the company and teamwork), and also theoretically contributes to knowledge. The findings of this study support the research of Sarasvathy (2001), in which all five dimensions of effectiveness were partly related to the decisions of entrepreneurs and especially internationalization (Prashantham, Kumar, Bhagavatula, & Sarasvathy, 2018; Sarasvathy, 2001). Although the work of entrepreneurs focuses on the "cost-effectiveness principle" when deciding to exit the international market, other aspects of effectiveness are more evident in other areas of planned internationalization. This suggests that specific elements of decision-making effectiveness can be identified in particular contexts. Current research, based on existing studies that include rapidly internationalized companies (Crick & Crick, 2018), examines respondents' reflections on their initial behavior. The findings indicate that due to risk and reward (Sarasvathy, 2001), all stakeholders are driven to create strategies based on "what they know" and "who knows" in response to their unpredictable environment. In this study, all interviewees have a comprehensive understanding of what they want and why they want it, but they often accept unplanned or unexpected orders without sufficient preparation (Mirvahedi & Morrish, 2017). To facilitate the growth of start-ups, all interviewees recognize the need for a balance between risk and reward, while also considering cost-effectiveness, and they quickly expand into small international markets through exports.

For instance, international experience offers various potential rewards, including revenue streams, initial credit, international management knowledge, and references for customers in other markets (Bell, Crick, & Young, 2004; Crick & Crick, 2014, 2018; Crick, 2016). Achieving these rewards requires components such as company growth, risk-taking, capacity assessment, opportunity awareness, and teamwork within the organization. In this context, Ruzzier et al. (2007) demonstrated that internationalization is directly linked to the inclination to internationalize, managerial knowledge, and international business skills (Ruzzier, Antoncic, Hisrich, & Konecnik, 2007). Therefore, when entering the international market, it is crucial to consider the different dimensions of "cost-effectiveness," encompassing both financial aspects (e.g., profit/loss) and non-financial aspects (e.g., credit). Other findings from this study indicate that stakeholders engaged in entrepreneurial marketing behavior to identify potential rewards (Morris, Schindehutte, & Laforge, 2002). They demonstrated active, risky, and innovative behavior, as well as engagement in opportunistic behavior and resource management. Thus, Morris et al. (2002) argued that all dimensions of entrepreneurial marketing need to be considered, except for a few. Ultimately, decision makers must carefully evaluate the market and establish negative connections in small, well-known markets where people frequently interact. The timing of internationalization may be unexpected, but it is not solely based on receiving an order. Although the existing marketing competency literature emphasizes extensive data collection, it is important to note that extensive data collection can be costly and time-consuming for start-ups with limited resources (Blank, 2013).

The research findings on the impact of risk-taking are consistent with the findings of Dehkordi, Nankali, & Jame (2017). Their research also revealed that risk-taking, as one of the dimensions of entrepreneurship, has a positive effect on business internationalization. It was also discovered that all the interviewees and the presence of all surveyed companies in the internationalization space are considered a sustainable competitive advantage at the international level. Consequently, companies that offer products at or below the level of foreign competitors are unable to establish an international presence and are therefore excluded from the international market (Sua′rez-Ortega & Alamo-Vera, 2005). They demonstrated that the inclination to export has a positive relationship with the company's competitive position in the development of new products and managerial perception, which is one of the company's advantages. Based on previous research, research has been conducted to demonstrate that internationalization is not always a step forward, considering the future goals of the interviewees and the impact of the decision (Bernini, Du, & Love, 2016; Yayla, Yeniyurt, Uslay, & Cavusgil, 2018). There is relatively limited research on the ongoing internationalization behavior of companies. The present study indicates that the "rest time" of shareholders after deciding to exit the international market reflects the way forward.

KEYWORDS

- **data collection**
- **decision making**
- **effectiveness**
- **entrepreneurial marketing**
- **international market**
- **overseas sales**
- **rapid internationalization**
- **re-internationalization**
- **shareholders**

REFERENCES

Ahmadi, H., & O'Cass, A., (2016). The role of entrepreneurial marketing in new technology ventures first product commercialization. *Journal of Strategic Marketing, 24*(1), 47–60.

Andersson, S., (2011). International entrepreneurship, born globals and the theory of effectuation. *Journal of Small Business and Enterprise Development, 18*(3), 627–643.

Arend, R. J., Sarooghi, H., & Burkemper, A., (2015). Effectuation as ineffectual? Applying the 3E theory-assessment framework to a proposed new theory of entrepreneurship. *Academy of Management Review, 40*(4), 630–651.

Azar, A., Sadeghi, A., & Kordnaeij, A., (2012). Prioritization of factors affecting the success of small and medium businesses in the superior technology area – fuzzy network analysis process approach. *Entrepreneurship Development Quarterly, 5*(12), 165–184.

Becherer, R. C., Helms, M. M., & McDonald, J. P., (2012). The effect of entrepreneurial marketing on outcome goals in SMEs. *New England Journal of Entrepreneurship, 15*(1), 7–18.

Bell, J., Crick, D., & Young, S., (2004). Small firm internationalization and business strategy: An exploratory study of 'knowledge-intensive' and 'traditional' manufacturing firms in the UK. *International Small Business Journal, 22*(1), 23–56.

Bell, J., McNaughton, R. B., Young, S., & Crick, D., (2003). Towards an integrative model of small firm internationalization. *Journal of International Entrepreneurship, 1*(4), 339–362.
Bernini, M., Du, J., & Love, J. H., (2016). Explaining intermittent exporting: Exit and conditional re-entry in export markets. *Journal of International Business Studies, 47*(9), 1058–1076.
Beverland, M., & Lindgreen, A., (2010). What makes a good case study? A positivist review of qualitative case research published in industrial marketing management in 1971–2006. *Industrial Marketing Management, 39*(1), 56–63.
Bjerke, B., & Hultman, C., (2002). *Entrepreneurial Marketing: The Growth of Small Firms in the New Economic Era.* Cheltenham: Edward Elgar.
Blank, S., (2013). Why the lean start-up changes everything. *Harvard Business Review, 91*(5), 64–68.
Collinson, E., & Shaw, E., (2001). Entrepreneurial marketing – a historical perspective on development and practice. *Management Decision, 39*(9), 761–766.
Crick, D., & Crick, J., (2014). The internationalization strategies of rapidly internationalizing high-tech SMES: Planned and unplanned activities. *European Business Review, 26*(5), 421–448.
Crick, D., (2016). The first export order: A marketing innovation revisited. *Journal of Strategic Marketing, 24*(2), 77–89.
Crick, J. M., & Crick, D., (2018). 'Lean start-up' practices: Initial internationalization and evolving business models. In: Leonidou, C. S. L. C., (ed.), *Advances in Global Marketing: A Research Anthology* (pp. 37–58). Cham: Springe.
Crick, J. M., & Crick, D., (2018b). Angel investors' predictive and control funding criteria: The importance of evolving business models. *Journal of Research in Marketing and Entrepreneurship, 20*(1), 34–56.
Dashti, M., (2009). *Identifing th Effective Factors on Internationalization of Industrial SMEs in Isfehan Province.* Isfehan: Master thesis. University of Isfehan.
Dehkordi, M. A., Nankali, A., & Jame, B. M., (2017). Investigating the impact of entrepreneurial orientation on internationalization of companies in glass and crystal industry by mediating role of organizational culture. *Journal of Entrepreneurship Develop.*
Gabrielsson, P., & Gabrielsson, M., (2013). A dynamic model of growth phases and survival in international business-to-business new ventures: The moderating effect of decision-making logic. *Industrial Marketing Management, 42*(8), 1357–1373.
Galkina, T., & Chetty, S., (2015). Effectuation and networking in internationalizing SMEs. *Management International Review, 55*(5), 647–676.
Gassmann, O., Frankenberger, K., & Csik, M., (2014). *The Business Model Navigator.* Harlow: FT Publishing.
Gilmore, A., (2010). Reflections on methodologies for research at the marketing/entrepreneurship interface. *Journal of Research in Marketing and Entrepreneurship, 12*(1), 11–20.
Hansen, D. J., & Eggers, F., (2010). The marketing/entrepreneurship interface: A report on the "Charleston Summit. *Journal of Research in Marketing and Entrepreneurship, 12*(1), 42–53.
Hewerdine, L. J., Rumyantseva, M., & Welch, C., (2014). Resource scavenging. Another dimension of the internationalization pattern of high-tech SMEs. *International Marketing Review, 31*(3), 237–258.

Hills, G. E., & Hultman, C. M., (2011). Academic roots: The past and present of entrepreneurial marketing. *Journal of Small Business & Entrepreneurship, 24*(1), 1–10.
Hills, G. E., Hultman, C. M., & Miles, M. P., (2008). The evolution and development of entrepreneurial marketing. *Journal of Small Business Management, 46*(1), 99–112.
Ibeh, K., Jones, M. V., & Kuivalainen, O., (2018). Consolidating and advancing knowledge on the post-entry performance of international new ventures. *International Small Business Journal, 36*(7), 741–757.
Ionita, D., (2012). Entrepreneurial marketing: A new approach for challenging times. *Management and Marketing, 7*(1), 131–150.
Johanson, J., & Vahlne, J. E., (1977). the internationalization process of the firm – a model of knowledge development and increasing foreign market commitment. *Journal of International Business Studies, 8*(1), 23–32.
Jones, M. V., & Coviello, N. E., (2005). Internationalization: Conceptualizing an entrepreneurial process of behavior in time. *Journal of International Business Studie, 36*(3), 284–303.
Jones, M. V., Coviello, N. E., & Tang, Y. K., (2011). International entrepreneurship research (1989–2009): A domain ontology and thematic analysis. *Journal of Business Venturing, 26*(6), 632–659.
Jones, R., & Rowley, J., (2011). Entrepreneurial marketing in small businesses: A conceptual exploration. *International Small Business Journal, 29*(1), 25–36.
Khayatian, M. S., Tabatabayan, S. H., Amiri, M., & Eliassy, M., (2015). Content analysis of the characteristics of knowledge-based companies. *Organizational Resource Management Research, 5*(2), 47–21. In Persian.
Lincoln, Y. S., & Guba, E. G., (2000). Paradigmatic controversies, contradictions, and emerging confluences. In: Denzin, N. K., & Lincoln, Y. S., (eds.), *The Handbook of Qualitative Research*. Beverly Hills, CA: Sage.
Matsuno, K., Mentzer, J., & Ozsomer, A., (2002). The effects of entrepreneurial proclivity and market orientation on business performance. *Journal of Marketing, 66*(3), 18–32.
McKelvie, A., Haynie, J. M., & Gustavsson, V., (2011). Unpacking the uncertainty construct: Implications for entrepreneurial action. *Journal of Business Venturing, 26*(3), 273–292.
Miles, M. P., & Darroch, J., (2006). Large firms, entrepreneurial marketing processes, and the cycle of competitive advantage. *European Journal of Marketing, 40*(5, 6), 485–501.
Miles, M., Gilmore, A., Harrigan, P., Lewis, G., & Sethna, Z., (2014). Exploring entrepreneurial marketing. *Journal of Strategic Marketing, 23*(2), 94–111.
Mirvahedi, S., & Morrish, S., (2017). The role of serendipity in opportunity exploration. *Journal of Research in Marketing and Entrepreneurship, 19*(2), 105–124.
Morris, M. H., Schindehutte, M., & Laforge, R. W., (2002). Entrepreneurial marketing: A construct for integrating emerging entrepreneurship and marketing perspectives. *Journal of Marketing Theory and Practice, 10*(4), 1–19.
Morrish, S. C., (2011). Entrepreneurial marketing: A strategy for the twenty-first century? *Journal of Research in Marketing and Entrepreneurship, 13*(2), 182–200.
Morrow, S. L., (2005). Quality and trustworthiness in qualitative research in counseling psychology. *Journal of Counseling Psychology, 52*(2), 250–260.
Prashantham, S., Kumar, K., Bhagavatula, S., & Sarasvathy, S. D., (2018). Effectuation, network-building, and internationalization speed. *International Small Business Journal*, https://doi.org/10.1177/0266242618796145.

Read, S., Sarasvathy, S. D., Dew, N., & Wiltbank, R., (2016). Response to Arend, Sarooghi and Burkemper: Co-creating effectual entrepreneurship research. *Academy of Management Review*, 528–536.

Ries, E., (2011). *The Lean Start-Up*. London: London: Pengui.

Roundy, P. T., (2017). "Doing well" while serving customers. Charting the social entrepreneurship and marketing interface. *Journal of Research in Marketing and Entrepreneurship, 19*(2), 105–124.

Ruzzier, M., Antoncic, B., Hisrich, R., & Konecnik, M., (2007). Human capital and SME internationalization: A structural equation modeling study. *Canadian Journal of Administrative Sciences, 24*(3), 15–29.

Sarasvathy, S. D., (2001). Causation and effectuation: Toward a theoretical shift from economic inevitability to entrepreneurial contingency. *Academy of Management Review, 26*(2), 243–263.

Saunders, M. N., Lewis, P., & Thornhill, A., (2009). *Research Methods for Business Students*. London: London: Pearson.

Schmid, J., (2012). *Entrepreneurial Marketing – Often Described, Rarely Measured.* A proposal for an operationalization of entrepreneurial marketing in SMEs based on established frameworks in the literature. Academy of Marketing Conference. Southampton.

Sinkovics, R. R., & Alfoldi, E. A., (2012). Progressive focusing and trustworthiness in qualitative research: The enabling role of computer-assisted qualitative data analysis software (CAQDAS). *Management International Review, 52*(6), 817–845.

Storey, D. J., (1994). *Understanding the Small Business Sector.* London: London: Routledge.

Sua´rez-Ortega, S., & Alamo-Vera, F., (2005). SMES' internationalization: Firms and managerial factors. *International Journal of Entrepreneurial Behavior & Research, 11*(4), 258–279.

Timmermans, S., & Tavory, I., (2012). Theory construction in qualitative research from grounded theory to abductive analysis. *Sociological Theory, 30*(3), 167–186.

Turner, C., (2012). De-internationalization: Towards a co-evolutionary framework. *European Business Review, 24*(2), 92–105.

Welch, C., & Welch, L. S., (2009). Re-internationalization: Exploration and conceptualization. *International Business Review, 18*(6), 567–577.

Welch, C., Piekkari, R., Plakoyiannaki, E., & Paavilainen-Mantymaki, E., (2011). Theorizing from case studies: Towards a pluralist future for international business research. *Journal of International Business Studies, 42*(5), 740–762.

Welch, L. S., & Wiedersheim-Paul, F., (1980). Initial exports – A marketing failure. *Journal of Management Studies, 17*(3), 333–344.

Woodside, A. G., Pattinson, H. M., & Miller, K. E., (2005). Advancing hermeneutic research for interpreting interfirm new product development. *Journal of Business & Industrial Marketing, 20*(7), 364–379.

Yayla, S., Yeniyurt, S., Uslay, C., & Cavusgil, E., (2018). The role of market orientation, relational capital, and internationalization speed in foreign market exit and re-entry decisions under turbulent conditions. *International Business Review, 27*(6), 11.

Zhu, Z., & Matsuno, K., (2016). Entrepreneurial proclivity: Its environmental conditions and growth consequences. *Journal of Strategic Marketing, 24*(1), 20–33.

CHAPTER 4

Cryptocurrency: The Revolution and Trend of Bitcoin in the Recent Scenario and Its Future Challenges and Opportunities

PRIYAJIT RAY

Research Scholar, Department of Commerce, University of Calcutta, Kolkata, West Bengal, India

ABSTRACT

Bitcoin is one of the most profound cryptocurrencies of the 21st century. Thanks to advancements in technological development, it has become possible and effective to invest in such coins from anywhere around the world. Since its inception, Bitcoin has gained considerable momentum and popularity among investors, despite facing legal issues. While there are several cryptocurrencies globally, Bitcoin remains the most popular, with many global investors financially dependent on it. Even a slight change in its value can have a significant impact on them. Over the years, the value of Bitcoin and its market capitalization have risen, attracting global investors to this digital currency.

Therefore, a study has been conducted to analyze the growth, trend, and importance of Bitcoin, as well as its challenges and possible opportunities in the current scenario and the future for its sustainability. The aim is to provide insights that can benefit investors in the economy. The researcher utilized a line chart and a simple linear regression model to examine the trend and growth of Bitcoin's value in recent terms. The findings of the study clearly indicate that there is growth in the value of Bitcoin, accompanied by a fluctuating trend. Based on these findings, suitable conclusions have been

Contemporary Digital Transformation and Organizational Effectiveness in Business 4.0.
Nilanjan Ray, Sudin Bag, & Tapas Kumar Chatterjee (Eds.)

drawn by the researcher to assist investors in making informed decisions regarding such currencies.

4.1 INTRODUCTION

Cryptocurrencies are virtual or digital currencies that take the form of coins or tokens. They are used as a medium of exchange and are secured by cryptography. They have gained significant popularity in today's era, thanks in part to the rapid development of information and communication technologies. These technologies have enabled many daily activities to be conducted online, offering greater flexibility and effectiveness in the long run.

Over the years, cryptocurrencies have been increasingly accepted on a global scale, as evidenced by the rise in their value and trading volume. Currently, there are more than 6,500 cryptocurrencies operating globally, but bitcoins have gained the most popularity since their inception just over a decade ago. However, bitcoins are also known for their high volatility, which has both excited and dissatisfied investors due to sudden price changes. It is important to note that the cryptocurrency market is a zero-sum game, meaning one investor's profit translates to another investor's loss or missed opportunity.

Bitcoin is one of the primary cryptocurrencies globally that can be stored and traded through the internet. It operates on an open-source financial network that is decentralized and based on a peer-to-peer digital system, operating globally without geographical boundaries. Since its inception, bitcoin has experienced exponential growth in the market, highlighting the growing interest of investors in this decentralized currency. While bitcoins are considered legal and free to invest in some countries, in India, it is not legal due to the lack of authorization and regulation of virtual coins, as well as the absence of a comprehensive law to govern bitcoin trading and ensure its legality. Therefore, it is necessary for the Indian government to enact a comprehensive law as soon as possible, as there are ongoing debates regarding the usage of bitcoins and even considerations of banning it in certain countries. However, there are investors who consider these coins invaluable with great value, making bitcoin one of the most profound revolutions of recent times.

4.2 REVIEW OF LITERATURE

Several pieces of literature were surveyed in the relevant field to identify research gaps and fulfill the objectives of the research study.

Sontakke & Ghaisas (2017) analyzed cryptocurrencies with blockchain technology, and various asset classes were also examined based on real-time and historical price data. The major findings of the study indicate that the demand and supply model drives cryptocurrency prices across different asset classes. The researchers further explained that forecasting cryptocurrency prices is relatively challenging, which poses difficulties for both the government and investors.

Malik (2020) conducted a study focusing on the drivers of bitcoin prices in the Indian macroeconomic environment. The researcher empirically analyzed the determinants of bitcoin prices in India from 2017 to 2019, considering a period of 156 weeks using the Vector Error Correction Model. The findings of the study suggest that the prices of gold, bitcoin supply, money supply, and the trade volume of bitcoin significantly impact the deviations of bitcoin prices from the equilibrium price.

Hung (2021) aimed to analyze the relationship between the Bitcoin market and conventional asset classes in our country using correlation and Granger causality tests. The researchers found evidence of a unidirectional connection between Bitcoin and other assets in India. Additionally, they discovered a significant interrelationship between Bitcoin and key financial asset returns at low, medium, and high frequencies.

Naresh & Ananda (2021) employed the Granger causality test to investigate the causal relationship between Bitcoin prices and the US-rupee dollar exchange rate during the pandemic. The researchers also examined the price behavior of Bitcoin and the rupee-dollar exchange rate. They found a unidirectional causality, where the rupee-US dollar exchange rate influenced the prices of Bitcoin in the Indian market during the pandemic. Furthermore, the rupee-dollar exchange rate had a significant impact on the prices of Bitcoin in our country.

4.3 RESEARCH GAP

The researcher has found that there is a lack of any definite research work undertaken yet to gain in-depth knowledge regarding the trend and growth of bitcoins. This knowledge would be beneficial for investors and to understand the future challenges and opportunities in the current scenario. The researcher identified this gap through an extensive review of literature in the concerned area. The researcher considers this as a major research gap with socioeconomic significance for the economy. The major avenue of investment lies with the investors, and therefore the researchers choose to fulfill the research gap using secondary data and information. The study aims to bridge the gap through the defined research objectives and raised research questions by conducting explorative, analytical, and empirical research work.

4.4 RESEARCH OBJECTIVES

The study focuses on the following research objectives based on the identified research gap:

- To study the growth and importance of bitcoins for investors and society as a whole.
- To analyze the trend of bitcoins in the recent scenario to predict its future performance.
- To understand the challenges and opportunities of Bitcoins for its sustainability in an economy.

4.5 RESEARCH QUESTIONS

The researcher has raised the following research questions based on the research objectives of the study:

- Are bitcoins important to investors and society in general?
- Does the value of bitcoins follow any specific trend that can be used to predict its future performance?
- What are the challenges and potential opportunities of bitcoins in the current scenario?

4.6 RESEARCH METHODOLOGY

The present study is of a descriptive, exploratory, and empirical nature, primarily relying on secondary sources of information. The exploratory and descriptive analysis of the available literature has been gathered from various articles, journals, news reports, bulletins, research works, published information, and other relevant reports in the field. These sources have been utilized to fulfill the research objectives. Additionally, various websites have been visited to address the research questions and explore the legal aspects of bitcoins, aiming to generate better prospects for potential investors and ensure the sustainability of such cryptocurrencies.

For the empirical part of the study, data pertaining to bitcoins has been collected and analyzed from the Nasdaq website. This data has been used to understand the growth and analyze the trend of bitcoins, enabling predictions about the future performance of these cryptocurrencies. The data covers a period of 10 quarters, from March 31, 2019, to September 30, 2021, which was determined based on the researcher's rationale. This timeframe was deemed sufficient to assess the growth and trend of bitcoin values.

Initially, the researcher collected relevant data and information on various aspects of bitcoins and their recent evolution. After data collection, the information was presented and analyzed using line graphs to comprehend the trend and growth of bitcoins based on their price and value. Apart from the line graph, a linear regression model has also been applied during the study period to determine if there is any trend in the dataset over time. This allows for testing the formulated hypothesis of the study. In the simple linear regression model, the dependent variable considered is the quarterly closing price value of bitcoins, while the independent variable is time, represented as the increase in the number of quarters. Considering that cryptocurrencies serve as a source of finance for many investors, the available and relevant data has been thoroughly and conceptually analyzed. The data has been presented and analyzed using Microsoft Excel and IBM SPSS software in this research work. Based on the results obtained from the study, suitable findings and logical conclusions have been drawn by the researcher through extensive data analysis. This helps in understanding the relevance of bitcoins among investors for overall growth and sustainable development of individuals and society as a whole.

4.7 RESEARCH HYPOTHESIS

Based on the importance of the study, the researcher has tested the following formulated hypothesis in order to answer its research questions:

- $\mathbf{H_{01}}$**:** There is no significant trend in the quarterly closing price value of bitcoins for the period of the study.
- $\mathbf{H_{11}}$**:** There is a significant trend in the quarterly closing price value of bitcoins for the period of the study.

4.8 DATA PRESENTATION AND ANALYSIS

Bitcoin represents the most profound revolution of the recent era. It operates virtually and electronically in the age of digital innovation and development. Furthermore, it is considered an electronic cash system, as the digital currency is based on a peer-to-peer payment system. While there are several other cryptocurrencies being operated in the global market, bitcoins have surpassed great heights in terms of value and have attracted investors to invest more in cryptocurrencies. Bitcoins are now considered the latest class of assets by investors, making their mark in the investment landscape. Although there are more than 5,000 cryptocurrencies being operated globally, bitcoin remains the most popular one, with many people unaware of other currencies. It is no secret that the growth of bitcoins has risen considerably, but its operations and legality remain a major concern for global investors, raising questions about its sustainability for the future.

4.8.1 GROWTH AND IMPORTANCE OF BITCOINS

The global cryptocurrencies, and bitcoins in particular, have experienced incredible growth since their inception. They have become a constant source of investment for many global investors. Over the years, bitcoins have evolved into a store of value. Although there have been many digital currencies since the 1980s, investors have primarily recognized cryptocurrencies through the lens of bitcoins. Despite experiencing sudden falls and rises since their introduction, the growth of bitcoins without any promotion or fanfare is truly remarkable. This highlights the importance of such cryptocurrencies for investors (Table 4.1).

- In the year 2011, the value of bitcoins rose at a pace of 3,200% within a period of just three months. The coins, which were valued at $1 in April, shot up considerably and reached a value of $32 in June 2011.
- There was also a significant rise in the value of bitcoins in 2013. The coins, which were valued at $13.40 in January, increased to $220 in April, and then soared to $1,156 in December 2013.
- In 2017, there was a staggering increase in the price of bitcoins. Its value showed remarkable growth from $975 in March to $20,089 in December.

- In 2020 as well, there was a considerable rise in the value of bitcoins. The cryptocurrency was valued at $7,200 in January and then increased to $18,353 in November.
- The price of bitcoins, which was valued at $24,000 in December 2020, rose to $40,000 in January 2021. It then reached its peak, hitting an all-time high of $64,000 in April 2021.
- While there were fluctuations in bitcoin prices globally, the overall rise in its value is incredible. Despite occasional drops, the rapid pace of its ascent highlights the importance of digital currencies in the investment field.
- The importance of bitcoins is also growing considering the increase in their value. Many investors are turning to digital currency as a source of income for their livelihood.
- The growing importance of bitcoins is also attributed to their invaluable features, such as greater scalability and security, lower transaction costs, and speedy processing. These factors make bitcoins more attractive to investors compared to other cryptocurrencies.

TABLE 4.1 Quarterly Closing Price Value of Bitcoins During the Period of the Study

SL. No.	Date	Bitcoin Quarterly Closing Price Value
1.	31st March 2019	3974.4
2.	30th June 2019	11074.2
3.	30th September 2019	8275.9
4.	31st December 2019	7222
5.	31st March 2020	6346.7
6.	30th June 2020	9124.2
7.	30th September 2020	10801.9
8.	31st December 2020	29238.8
9.	31st March 2021	58995.8
10.	30th June 2021	34202.6
11.	30th September 2021	43467.3

Source: Retrieved from: https://www.nasdaq.com/market-activity/cryptocurrency/btc/historical.

From Figure 4.1, it is evident that there is both a rising and fluctuating trend in the quarterly closing price valuation of bitcoins during the study period. Initially, when bitcoins were introduced, their valuation started slowly and then experienced moderate fluctuations. However, in the last quarter of December 2020 and the first quarter of January 2021, there was a significant increase in valuation at an alarming rate. Nevertheless, the valuation of bitcoins continued to fluctuate, making it unpredictable in nature. Overall, there is observed growth in the valuation of bitcoins during the study period, but the fluctuations between months make it challenging for investors to predict its future performance and rely on bitcoins as a reliable investment.

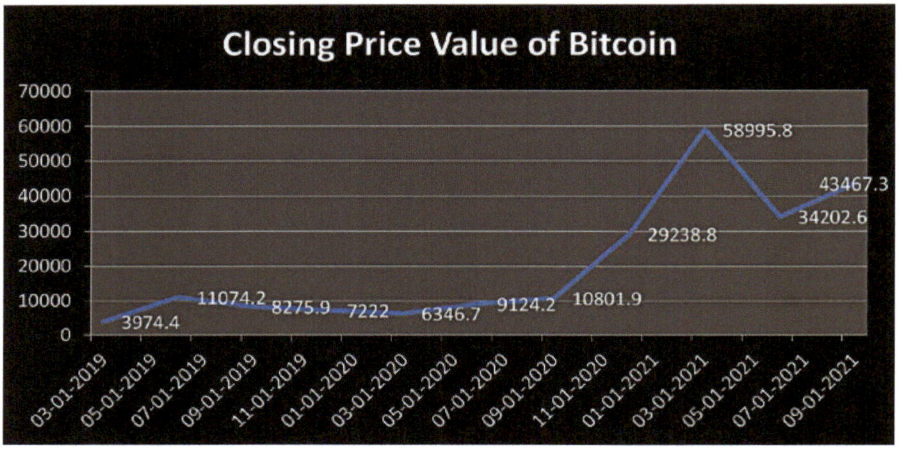

FIGURE 4.1 Quarterly closing price value of bitcoins during the period of the study.

Source: Researcher's presentation using MS excel.

To further analyze the price and valuation of bitcoins in the recent scenario for the period of the study, the following linear regression model has been developed and its equation is shown below:

$$BQCPV = a0 + bt + \varepsilon$$

where; 'BQCPV' is the bitcoin quarterly closing price value; 'a0' and 'b' are parameters indicating intercept and coefficient; 't' is the time factor for the study period; and 'ε' is the error term.

From Table 4.2, it is evident that the correlation coefficient (R) between the dependent variable, quarterly closing price value of Bitcoin, and time exhibits a significant linear association at 0.804, indicating a high correlation between the variables. The R square value of 0.647 indicates that more than 65% of the quarterly closing price value of Bitcoin can be explained by the independent variable, time, during the study period. The Durbin Watson value of 1.541 suggests the absence of first-order autocorrelation in the model, and the VIF value of 1 indicates no multicollinearity. The F value of 16.467, significant at a 5% level of significance, indicates that the regression model is a reliable predictor. Additionally, the unstandardized regression coefficient (beta) value of 4460.242 in Table 4.2 signifies a positive and significant trend in the quarterly closing price value of Bitcoin at a 5% level of significance.

TABLE 4.2 Trend Analysis of Quarterly Closing Price Value of Bitcoins During the Period of the Study[a,b]

R	R Square	Durbin-Watson	F	Sig.	Unstandardized Coefficient (B)	VIF
0.804	0.647	1.541	16.467	0.003	4460.242	1.000

[a] Predictors: (Constant), time;
[b] Dependent variable: BQCPV.

Source: Researcher's computation using SPSS.

4.8.2 CHALLENGES AND OPPORTUNITIES OF BITCOINS

Bitcoins have been widely adopted by people all over the world, with over 100 million investors globally. However, it is important to note that bitcoins are decentralized and not legal in India. Despite its rapid rise in value, potential investors may hesitate due to the various challenges associated with bitcoins.

One of the most significant challenges is the legality of cryptocurrencies. In India, the trading, creation, and usage of bitcoins are not authorized, making it risky to invest in them. Investors who choose to invest in bitcoins often find themselves under the scrutiny of law enforcement agencies due to the nature of this digital currency. Additionally, there is a taxation issue surrounding bitcoins in our economy. Since cryptocurrencies lack authorization and legal status, there is an ongoing debate regarding the taxation of bitcoin trading.

Another challenge is the varying legal status of bitcoins across different countries. While it may be legal in some countries, it is restricted in others. This can lead to exchange rate discrepancies when transferring bitcoins between countries.

There are several benefits and opportunities of bitcoins, ranging from their application in the healthcare sector to the real estate sector. Bitcoin currency can be utilized for the automatic disbursement of healthcare payments and small payments under various health plans. Bitcoins can also be effectively employed in the real estate sector to address the issue of bureaucracy and reduce processing time. This revolutionary cryptocurrency can also be used for storing records related to intellectual property, facilitating transparent transactions for licensing and transfers. Bitcoins can be utilized for contractual payment obligations in the legal services sector, resulting in compliance directives and reduced monitoring resources. They can also provide solutions for revenue sharing and rights management among various media enterprises. Additionally, bitcoins can enhance the accessibility of insurance and microinsurance, enabling micro-payments based on data availability from connected mechanisms. With the recent advancements in blockchain technology, bitcoins can save time and money across different industries, offering numerous opportunities for automation and self-execution to achieve desired objectives in every sector.

4.9 DISCUSSION AND CONCLUSION

Based on the data presentation and analysis of the study, it is evident that there are significant results regarding the formulated testable hypothesis. The study aims to analyze the trend in the quarterly closing price value of bitcoins in order to predict its future performance. The findings indicate the following:

- $\mathbf{H}_{01}$ is rejected at a 5% level of significance, suggesting the presence of a significant trend in the quarterly closing price value of bitcoins during the study period.

Referring to Table 4.2 of the study, it is clear that the Null Hypothesis is rejected due to a p-value of less than 0.05. The study's figure also demonstrates that the value of bitcoins has increased over time, albeit with a fluctuating trend. This fluctuation makes it challenging for investors to predict its value accurately. However, there is an overall upward trend, making bitcoins an attractive investment avenue. It is important to note that the legality of such currencies remains a major concern.

Cryptocurrencies, particularly bitcoins, have revolutionized the current financial landscape. Their popularity stems from the significant rise in value since their inception. Despite the fluctuating nature of bitcoins, they face serious challenges in many countries, which is a major concern for investors and economies as a whole.

The growth and rise in value of bitcoins have been remarkable since their introduction, as evidenced by the analysis of the study. While authorization and legality pose significant challenges for this cryptocurrency, there are numerous opportunities for bitcoins in various fields, which are likely to ensure the sustainability of such digital currencies in the free-flowing market. Initially, bitcoins were used as a currency for daily transactions and trading, but their volumes remain a fraction of their mainstream counterparts, and they never truly gained traction as a medium of daily transactions. Bitcoin is not just a digital currency; it has evolved into a store of value, with investors choosing to buy and hold it for long periods of time rather than using it for daily transactions. This shift in investor mindset highlights the importance of bitcoins. Despite being decentralized and borderless, bitcoins represent a profound and peaceful revolution of the 21st century, with investors appreciating their scalability, quick processing, and lower transaction charges. While legal issues surround cryptocurrencies, it would be more beneficial to address these issues in a measured manner, involving representatives from all stakeholders, in order to develop standards and maximize benefits for society as a whole.

4.10 LIMITATIONS OF THE STUDY AND FUTURE RESEARCH SCOPE

The future researchers can conduct a research study based on collecting primary sources of data to understand the perception of investors and individuals. This is because the current research work has only been conducted using secondary sources of available data and information. Additionally, future researchers can also conduct a comparative study on bitcoins and other types of cryptocurrencies to obtain more concrete comparison results. The current study is limited to considering data from only a few quarters, so future researchers can consider a larger number of periods as a sample to obtain better and more accurate results regarding the trend and growth of bitcoins. Furthermore, future researchers can utilize more suitable and sophisticated statistical tools to gather more in-depth knowledge about bitcoins and their future predictability, for the benefit of investors.

KEYWORDS

- **bitcoins**
- **cryptocurrencies**
- **digital currency**
- **financial landscape**
- **growth**
- **healthcare**
- **statistical tools**
- **trend**

REFERENCES

Extance, A., (2015). The future of cryptocurrencies: Bitcoin and beyond. *Nature News, 526*(7571), 21.

Hung, N. T., (2021). Co movements between bitcoin and other asset classes in India. *Journal of Indian Business Research.*

Jani, S., (2017). *Scope for Bitcoins in India.* Retrieved from: Research Gate: www.researchgate.net/publication/321780780_Scope_for_Bitcoins_in-India (accessed on 09 March 2024).

Malik, S., (2020). Drivers of bitcoin prices: An empirical analysis of India. *Journal of Critical Reviews, 7*(14), 1252–1258.

Naresh, G., & Ananda, S., (2021). Bitcoin prices and rupee dollar exchange rates during COVID-19. *International Journal of Electronic Finance, 10*(3), 180–190.

National Association of Securities Dealers Automated Quotations. Retrieved from: https://www.nasdaq.com/market-activity/cryptocurrency/btc/historical (accessed on 09 March 2024).

Parashar, N., & Rasiwala, F., (2019). Bitcoin-asset or currency? User's perspective about cryptocurrencies. *IUP Journal of Management Research, 18*(1).

Rahman, A., & Dawood, A. K., (2019). Bitcoin and future of cryptocurrency. *Ushus Journal of Business Management, 18*(1), 61–66.

Singh, A. K., & Singh, K. V., (2018). Cryptocurrency in India-its effect and future on economy with special reference to bitcoin. *International Journal of Research in Economics and Social Sciences (IJRESS), 8*(3).

Sontakke, K. A., & Ghaisas, A., (2017). Cryptocurrencies: A developing asset class. *International Journal of Business Insights & Transformation, 10*(2).

CHAPTER 5

Analysis of Technology Acceptance Model in Understanding Higher Educational Institutions' Students' Attitude for Accepting Digital Education

AMRITA MAJUMDAR[1] and SUDIPTA MAJUMDAR[2]

[1]*Associate Professor, Amity Business School, Amity University, Kolkata, West Bengal, India*

[2]*Associate Professor, School of Business and Economics, Adamas University, Kolkata, West Bengal, India*

ABSTRACT

Educational institutions implement digital learning with multiple visions and objectives. The case arises more so for higher educational institutions. Therefore, when the COVID-19 pandemic reached India, all stakeholders of education, including students, parents, and educational administrators, were at a loss. Gradually, higher educational institutions started to transition to the world of digital education. Although digital education is slowly gaining popularity, very few studies have been conducted to understand students' attitudes towards accepting digital education. As a result, a study was conducted among university students to determine their attitude towards accepting digital education. The conceptual framework of this study is based on the technology acceptance model (TAM). The results of this study will assist educational administrators in implementing digital learning among students in a more effective manner.

Contemporary Digital Transformation and Organizational Effectiveness in Business 4.0.
Nilanjan Ray, Sudin Bag, & Tapas Kumar Chatterjee (Eds.)

5.1 INTRODUCTION

There has been a drastic change in the way education is delivered, particularly in Higher Educational Institutions. The COVID-19 pandemic has forced the education system to transition to digital platforms. As the new normal emerges, institutions have started adopting a blended mode of education delivery. This is especially true in the case of India, where the pre-COVID phase did not witness the evolution of online learning environments. There are barriers from both the perspectives of institutions and students. Changes in student characteristics, market conditions for educational delivery, and innovative technology itself are driving this development (Concannon, Flynn, & Campbell, 2005). However, various constraints, such as technical infrastructure, teacher effort, technological satisfaction, and graduate competency, limit the use of instructional technology in higher education (Surry, Ensminger, & Haab, 2005). Even several higher-level online educational institutions have faced challenges due to costly technology, poor choices, tough competition, and a lack of a business model (Elloumi, 2004). The delivery, efficiency, and acceptability of courses are just a few of the many digital learning tactics that many higher education institutions struggle to implement (Saadé, 2003).

Simply providing every imaginable course and striving to replicate the in-person classroom experience online cannot meet the demands of students and may lead to unforeseen failures (Kilmurray, 2003). Another issue with online learning is the persistent frustration of university students with regards to online education. This calls for further research on online education that focuses on the students (Hara, 2000). Given the growth and reliance on information systems, as well as the rapid integration of new-age technologies into learning, identifying the key factors for user acceptance of technology remains a crucial task (Yi & Hwang, 2003).

India fully utilizes ICT to promote the development of human resources (HRs) at all levels of education, with e-learning being considered one of the most significant choices for the modern knowledge-based society (Kim & Santiago, 2005). According to research by the Economist Intelligence Unit, India ranks as the fifth most prepared country in the world for e-learning (2003). Since the late 1990s, the majority of institutions have been offering partially or fully online e-learning courses. Currently, most offline universities either use e-learning or have a strategy in place for its implementation. Despite the quantitative growth of e-learning, the importance of quality evaluation in higher education in India is increasing (Lee, 2006). Additionally, there are still obstacles to the adoption of e-learning in universities and colleges (Leem & Lim, 2007).

Therefore, in order to effectively utilize an e-learning strategy to enhance learning, it is crucial for developers and providers of e-learning to possess a comprehensive understanding of how students perceive and respond to different aspects of e-learning (Koohang & Durante, 2003). Furthermore, by recognizing students' objectives and comprehending the factors that influence their perceptions of e-learning, academic administrators and managers can develop strategies to encourage more students to engage in this learning environment (Grandon, Alshare, & Kwan, 2005). Hence, conducting research that focuses on how learners perceive, feel about, and intend to use online learning is of utmost importance. However, there has been limited experimental research conducted to establish a connection between students' utilization of digital learning in higher education.

Drawing primarily on the technology acceptance model (TAM), this study presents an integrated theoretical framework of university students' attitudes towards adopting digital learning (TAM). The objectives of this study are to examine the relationships between university students' intentions to use digital learning and specific constructs, such as their attitudes, perceptions of the technology's usefulness and ease of use, self-efficacy in utilizing it, subjective norms, and system accessibility in relation to accepting e-learning, which will further contribute to the understanding of technology implementation.

5.2 LITERATURE REVIEW

The TAM, first proposed by Davis in 1986, is one of the well-known models associated with technology adoption and utilization. TAM has proven to be a valuable theoretical model for explaining and predicting user behavior in information technology (Legris, Ingham, & Collerette, 2003). According to Ajzen and Fishbein, TAM is considered a significant extension of the Theory of Reasoned Action (TRA) (1980). By modifying TRA, Davis (1989) and Davis, Bagozzi, & Warshaw (1989) introduced TAM as a theory to explain why users choose to adopt or reject information technology. TAM provides a framework for examining how environmental factors influence one's beliefs, attitudes, and intentions to use technology. TAM proposes two cognitive beliefs: perceived usefulness and perceived ease of use. TAM suggests that actual usage is influenced by these beliefs (Figure 5.1).

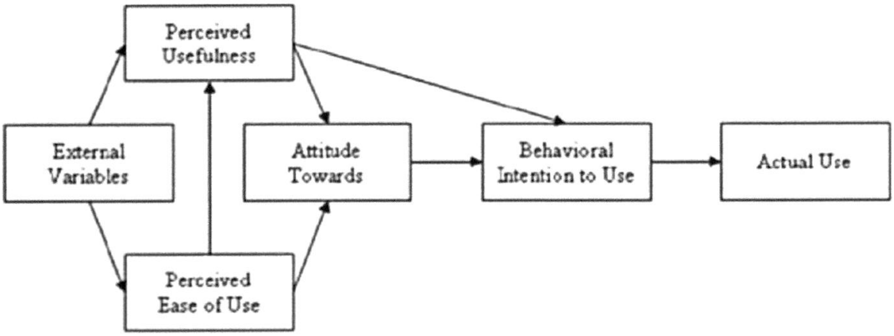

FIGURE 5.1 Technology acceptance model (TAM).

TAM appears to explain between 40% and 50% of consumer acceptability. TAM has undergone changes over the years. The TAM2 model incorporated social impact (subjective norm, voluntariness, and image) as well as cognitive instrumental processes (job relevance, output quality, and outcome demonstrability) to perceived usefulness and usage intentions. Both voluntary and obligatory testing environments were utilized for the new model. The findings provided significant support for TAM2 and accounted for 60% of user acceptance of this enhanced TAM (Venkatesh & Davis, 2000). In addition to TAM, this study employed TAM2 as the baseline model. TAM has been employed in numerous research studies to elucidate how individuals accept and utilize e-learning. According to Selim (2003), web-based learning must be examined in relation to TAM. He developed the course website acceptance model (CWAM) and employed structural equation modeling (SEM) techniques from the LISREL software to assess the connections between perceived usefulness, perceived usability, and intention to use among university students. He concluded that the model was consistent with the collected data and that the utility and usability of a course website were crucial factors in determining its acceptability and use. Perceived utility elucidates the extent to which a student perceives e-learning as beneficial for their academic performance. The objective of Venkatesh & Davis (1996) was to comprehend the factors contributing to perceived usability. They reached the conclusion that, although objective usability only had an impact on ease of use after actual use, computer self-efficacy influenced perceived ease of use. According to the argument presented by Grandon, Alshare, and Kwan in 2005, students' intentions were indirectly affected by their perception of the simplicity of using e-learning. Additionally, Mungania & Reio discovered a statistically significant correlation between dispositional barriers and self-efficacy in online learning in 2005. They argued that educational professionals should assess the learning styles of their students and seek strategies to enhance e-learning self-efficacy.

According to TAM2, one of the social impact factors known as subjective norm relates to the perceived social pressure to engage or not engage in the activity (Ajzen, 1991). To comprehend, justify, and predict system usage and acceptance behavior, it is crucial to understand how social variables influence the user's commitment to using the information system (Malhotra & Galletta, 1999). Subjective norm was found to be a key factor influencing university students' inclination to utilize e-learning, as indicated by a study conducted by Gradon, Alshare, & Kwan in 2005. In contrast, research conducted by Ndubisi in 2006 revealed that the desire of university students to utilize e-learning was not significantly affected by subjective norm. SEM, which identifies both direct and indirect effects, including spurious effects, can be employed to address such contradictions (Sobel, 1987).

Organizational context highlights the impact or support of an organization on an individual's usage of information technology, while social context refers to the social influence on an individual's acceptance of personal information technology use. Thong, Hong, & Tam (2002) identified relevance, system visibility, and system accessibility as factors within the organizational context. They argued that the organizational environment influences how valuable and user-friendly a digital library is perceived to be. Lin & Lu (2000) found that increased information accessibility leads to greater information utilization and a sense of ease of use.

Venkatesh & Davis (2000) expanded the model by introducing additional parameters, resulting in the TAM2 model. The factors that were added and influence the "perceived ease of use" and "perceived usefulness" are social influence determinants and cognitive instrumental determinants. Specifically, these factors include subjective norm, job relevance, output quality, result demonstrability, perceived ease of use, and perceived usefulness.

The TAM model has been widely used as a tool in various assessments related to information systems and is widely accepted by the scientific community due to its simplicity (Hsu & Chang, 2013). Researchers consider the results statistically reliable. It can be applied in different settings, such as the IT industry for assessing and researching information systems or new products, in distance learning to measure user acceptance, and specifically in platforms used for marketing and sales, as well as learning progress, such as online shopping and internet banking.

In research conducted in Korea, Park (2009) employed an expanded TAM model to examine the acceptance of e-learning courses among students who have taken at least one course at the nation's institution. The study aimed to investigate the factors influencing students' decision to accept e-learning courses. The findings revealed that self-efficacy had the greatest impact on behavioral intention, followed by social influence. This can be explained by the incentive theory, which suggests that students' influence plays a significant role in their utilization of online courses.

Another study conducted by Saadé, Nebebe, & Tan (2007) involved 362 Concordia University students in Canada who were enrolled in courses using a multimedia learning system. These students willingly participated in the study and completed a TAM questionnaire. The study utilized the Multimedia Acceptance Model (MAM) to assess students' satisfaction with the multimedia tool. The results demonstrated that "attitude" plays a crucial role in behavioral intentions, and "perceived usefulness" significantly influences attitude towards the use of MML. Additionally, a negative correlation was found between the "perceived utility" of the system and "behavioral intention," as well as between "perceived ease of use" and "attitude."

Furthermore, Dasgupta, Granger, & McGarry (2002) attempted to employ a courseware management platform to assess the user adoption of electronic collaboration technology. Around 62 students who had taken at least three courses and utilized this tool in one semester participated in the study using an extended TAM. According to the results, perceived usefulness had a positive effect on system utilization rather than having a negative influence. Additionally, system utilization could affect students' performance in the course.

Another comparable study based on TAM was conducted in Taiwan with the aim of examining Moodle's acceptability using an expanded TAM (Hsu & Chang, 2013). The perceived ease of use was introduced as an additional component in this research. Around 37 senior high school students and 47 college students participated in this study. The results indicated that perceived usefulness and attitude toward using Moodle were both influenced by perceived convenience, among other factors.

Behrens, Cranston, Jamieson, and Jones attempted to research the adoption of the "OASIS" system in 2005 (Online Assignment Submission, Infocom System). An extended TAM was employed in this study, and the sample of 112 students was divided into 94 "OASIS" users and 18 non-users. The results showed that, according to the TAM theory, perceived utility and perceived ease of use are good indicators of a system's success and can help explain why a system is accepted.

Another researcher, Liaw (2008), made an effort to examine the usefulness of the Blackboard e-learning system, the behavioral intentions of students, and their satisfaction with it. She also aimed to explain the reasons behind certain students' dissatisfaction. By utilizing an integrated TAM based on social cognitive theory and the theory of planned behavior, she collected responses from 424 university students. The results indicate a significant relationship between students' behavioral intentions to engage in e-learning and its effectiveness. Perceived self-efficacy was found to be the most important factor influencing students' satisfaction with the Blackboard e-learning system. Furthermore, perceived satisfaction and effectiveness may be influenced by students' behavioral intentions to engage in e-learning.

Additionally, Ngai, Poon, & Chan (2007) sought to explore the adoption of a web course tools (WebCT) by incorporating technical assistance as a possible component (WebCT). The survey involved 836 college students from seven universities in Hong Kong. According to the research, perceived utility and simplicity of use are the primary variables influencing students' attitudes towards using WebCT, with technical support having a significant impact on both. The results demonstrated the significance of perceived usability and utility in the relationship between technical support, attitude, and WebCT usage.

5.3 RESEARCH METHODOLOGY

The current study is an analytical research project aiming to identify students' attitudes towards accepting Digital Education. The respondents were selected from students in different higher educational institutions, including Colleges and Universities in Jharkhand, India. The dependent variable in this study is the attitude towards digital education, while the independent variables include self-efficacy in digital learning, subjective norm, system accessibility, perceived utility, and perceived usability. The responses were gathered from students at various higher education institutes in the Indian state of Jharkhand, with a total of 297 replies used in the analysis.

The primary data was collected through a structured questionnaire. The data was captured using Google Forms, which was circulated online through different professional and social networks, as well as emails among the students of HEIs in Jharkhand.

The questionnaire consists of two parts. The first part aims to understand the demographic aspects and the type of institutions to which the respondents belong. The second part focuses on understanding the students' attitudes towards accepting Digital Education, considering various independent variables such as digital-learning self-efficacy, subjective norm, system accessibility, perceived usefulness, and perceived ease of use.

The above variables were plotted based on a literature review and peer discussions. The students' perspective on Digital Education was framed by referring to the scale developed by Park (2009). Additionally, the scales by Han (2021); and Revyethi (2017) were used to frame the variables related to knowledge about pedagogical content and other parameters.

1. **Development of Questionnaire:** In this study, a structured questionnaire was utilized. The items for the structured questionnaire were gathered from the literature by Park (2009).
2. **Data Collection Method:** This employed in this study involved obtaining information from primary sources using a questionnaire as an instrument. The responses for the independent variable were collected using a five-point scale as follows: (1) strongly disagree (SD); (2) disagree (D); (3) neither agree nor disagree (NAD); (4) agree (A); and (5) strongly agree (SA), as proposed by Park (2009).

   The responses for dependent variable were collected using the following five-point scale (Park, 2009) – (1) very unhappy (VUH); (2) unhappy (UH); (3) neither unhappy nor happy (NUNH); (4) happy (H); and (5) very happy (VH).
3. **Analysis:** One-way ANOVA is used to investigate the measurable effect on the dependent variables.

5.4 ANALYSIS

5.4.1 DIGITAL LEARNING SELF-EFFICACY AND ATTITUDE TOWARDS DIGITAL EDUCATION

The psychographic variable being studied is digital learning self-efficacy. A one-way ANOVA is conducted to determine whether digital learning self-efficacy has a significant impact on attitudes towards Digital education. Data has been collected on the following items to assess self-efficacy towards Digital Learning.

- **DL1:** I feel confident finding information in the learning management system.

- **DL2:** I have the necessary skills for using a learning management system.

  The relevant portion of the SPSS output sheet is presented in Table 5.1 to infer whether there is any significant effect of digital learning self-efficacy on the Attitude towards Digital education.

TABLE 5.1 ANOVA Output for Digital Learning Self-Efficacy and Attitude Towards Digital Education[a,b]

Model		Sum of Squares	df	Mean Square	F	Sig.
1.	Regression	6.346	2	3.173	12.047	0.000[b]
	Residual	77.433	294	0.263		
	Total	83.779	296			

[a] Dependable variable: Attitude.
[b] Predictors: (Constant), DL2, DL1.

Source: SPSS output.

- **H_0:** Digital learning self-efficacy does not influence attitude towards digital education.

  The exact significance level (p-value) of the ANOVA is displayed in the 6th column (Sig.) of the aforementioned table. The significance level we have set is 5%, i.e., $\alpha = 0.05$ (based on existing research of a similar nature). The table indicates that the 'p' value is less than the 'α' value. Specifically, since $p = 0.000$ is less than $\alpha = 0.05$, the null hypothesis is rejected, and the alternative hypothesis is accepted. This implies that Digital Learning self-efficacy influences attitudes towards digital education.

5.4.2 SUBJECTIVE NORMS AND ATTITUDE TOWARDS DIGITAL EDUCATION

The psychographic variable being studied is subjective norms. A one-way ANOVA was conducted to determine whether Subjective Norms have a significant influence on the attitude towards using the Digital Learning system. Data has been collected on the following items to understand Subjective Norms:

- **SN1:** What digital learning stands for is important to me as a university student.

- **SN2:** I enjoy using digital learning based on the similarity of my values and society's values underlying its use.
- **SN3:** It is necessary for me to take digital learning courses in order to prepare for future job opportunities.

The relevant portion of the SPSS output sheet is presented in Table 5.2 to infer whether there is any significant effect of Subjective Norms on the attitude towards Digital education.

TABLE 5.2 ANOVA Output for Subjective Norms and Attitude Towards Digital Education[a,b]

Model		Sum of Squares	df	Mean Square	F	Sig.
1.	Regression	6.014	3	2.005	7.553	0.000[b]
	Residual	77.765	293	0.265		
	Total	83.779	296			

[a] Dependable variable: Attitude.
[b] Predictors: (Constant), SN3, SN2, SN1.
Source: SPSS output.

- **H_0:** Subjective norms does not influence attitude towards digital education.

The p-value for the ANOVA is shown in the 6th Col. (Sig.) of Table 5.2. We set the significance level at 5%, which is $\alpha = 0.05$ based on similar research. The table indicates that the p-value is less than the α-value. In fact, since $p = 0.000$ is less than $\alpha = 0.05$, we reject the null hypothesis and accept the alternative hypothesis. This means that Subjective Norms have an impact on attitudes towards digital education.

5.4.3 SYSTEM ACCESSIBILITY AND ATTITUDE TOWARDS DIGITAL EDUCATION

The psychographic variable being studied is system accessibility. A one-way ANOVA is conducted to determine whether system accessibility has a significant influence on the attitude towards using the digital learning system. Data has been collected on the following items to assess system accessibility.

- **SA1:** I have no difficulty accessing and using a learning management system at the university. The relevant portion of the SPSS output sheet is presented in Table 5.3 to infer whether there is any significant effect of system accessibility on the attitude towards digital education.

TABLE 5.3 ANOVA Output for System Accessibility and Attitude Towards Digital Education[a,b]

Model		Sum of Squares	df	Mean Square	F	Sig.
1.	Regression	3.487	1	3.487	12.811	0.000[b]
	Residual	80.292	295	0.272		
	Total	83.779	296			

[a] Dependable variable: Attitude.
[b] Predictors: (Constant), SA1.

Source: SPSS output.

- $\mathbf{H_0}$: System accessibility does not influence attitude towards digital education.

  The ANOVA table shows the p-value in the 6th Col. (Sig.). Our significance level is set at 5%, $\alpha = 0.05$ (based on similar research). The table indicates that the p-value is less than α. In fact, since $p = 0.000$ is less than $\alpha = 0.05$, we reject the null hypothesis and accept the alternative hypothesis. This means that system accessibility influences attitudes towards digital education.

5.4.4 PERCEIVED USEFULNESS AND ATTITUDE TOWARDS DIGITAL EDUCATION

The psychographic factors under investigation are the perceived utility. A one-way ANOVA has been conducted to determine whether system accessibility has a significant impact on the attitude towards utilizing the Digital Learning system. Data has been collected on the following items to comprehend the perceived usefulness.

- **PU1:** Digital learning would improve my learning performance.
- **PU2:** Digital learning would increase academic productivity.
- **PU3:** Digital learning could make it easier to study course content.

  The relevant portion of the SPSS output sheet is presented in Table 5.4 to infer whether there is any significant effect of Perceived usefulness on the Attitude towards Digital education.

TABLE 5.4 ANOVA Output for Perceived Usefulness and Attitude Towards Digital Education[a,b]

Model		Sum of Squares	df	Mean Square	F	Sig.
1.	Regression	16.169	3	5.390	23.358	0.000[b]
	Residual	16.170	293	0.231		
	Total	83.779	296			

[a] Dependable variable: Attitude.
[b] Predictors: (Constant), PU3, PU2, PU1.

Source: SPSS output.

- **H_0:** Perceived usefulness does not influence attitude towards digital education.

  The precise level of significance (p value) of the ANOVA is displayed in the 6th Column (Sig.) of the aforementioned table. The significance level established by us is 5%, namely $\alpha = 0.05$ (based on previous research of a similar nature). The table indicates that the 'p' value is lower than the 'α' value. In fact, since $p = 0.000$ is less than $\alpha = 0.05$, the null hypothesis is rejected, and the alternative hypothesis is accepted. This implies that perceived usefulness has an impact on attitude towards digital education.

5.4.5 PERCEIVED EASE OF USE AND ATTITUDE TOWARDS DIGITAL EDUCATION

The psychographic variables under investigation are the perceived ease of use. A one-way analysis of variance (ANOVA) was conducted to determine whether system accessibility has a significant impact on attitudes towards using the Digital Learning system. Data has been collected on the following items to gain insight into the perception of ease of use.

- **PE1:** I find learning management system easy to use.
- **PE2:** Learning how to use a learning management system is easy for me.
- **PE3:** It is easy to become skilled at using a learning management system.

  The relevant portion of the SPSS output sheet is presented in Table 5.5 to infer whether there is any significant effect of perceived ease of use on the attitude towards digital education.

TABLE 5.5 ANOVA Output for Perceived Ease of Use and Attitude Towards Digital Education[a,b]

Model		Sum of Squares	df	Mean Square	F	Sig.
1.	Regression	38.793	3	12.931	84.221	0.000[b]
	Residual	44.986	293	0.154		
	Total	83.779	296			

[a] Dependable variable: Attitude.
[b] Predictors: (Constant), PE3, PE2, PE1.

Source: SPSS output.

- **H_0:** The influence of perceived ease of use on attitude towards digital education is not observed.

  The precise level of significance (p value) of the analysis of variance (ANOVA) is prominently displayed in the sixth column (significance) of the aforementioned table. The significance level established by our research team is 5%, specifically $\alpha = 0.05$ (based on previous research of a similar nature). The table indicates that the 'p' value is lower than the 'α' value. In fact, since the p-value of 0.000 is less than the α-value of 0.05, the null hypothesis has not been accepted and the alternative hypothesis has been accepted. This implies that the perceived ease of use does indeed exert an influence on the attitude towards digital education.

5.5 CONCLUSION

As mentioned previously, the present study examines various psychographic variables, including digital-learning self-efficacy, subjective norm, system accessibility, perceived usefulness, and perceived ease of use, in relation to the attitude of students in Higher Educational Institutions towards digital education. The findings indicate that all of the aforementioned psychographic variables significantly influence the attitude of higher educational institution students towards digital education. This suggests that educational administrators should have a thorough understanding of Digital learning self-efficacy in order to develop a digital education framework for their institutions. Similarly, institutions can motivate and train students to utilize digital learning by emphasizing its benefits for their professional growth. Additionally, higher educational institutions should focus on subjective norms, ensuring the usefulness and user-friendliness of the system to alleviate any apprehensions students may have. Providing comprehensive system usage manuals, along with training and orientation sessions, can be beneficial. By considering these factors, higher educational institutions can foster a positive attitude towards digital education among students, as suggested by the study. Furthermore, this research can be expanded by surveying students' adaptability to digital education across different educational systems, particularly in the Indian context. Additionally, the study can be extended to other geographical locations and diverse social backgrounds.

KEYWORDS

- **COVID-19**
- **digital education**
- **digital transformation**
- **digital learning**
- **geographical locations**
- **higher education institute**
- **technology acceptance model**

REFERENCES

Ajzen, I., & Fishbein, M., (1980). *Understanding Attitudes and Predicting Social Behavior.* Englewood Cliffs, NJ: Prentice-Hall.

Bandura, A., (1994). Self-efficacy. In: Ramachaudran, V. S., (ed.), *Encyclopedia of Human Behavior* (pp. 71–81). New York, NY: Academic Press.

Bentler, P. M., & Bonnet, D. G., (1980). Significance tests and goodness-of-fit in the analysis of covariance structure. *Psychological Bulletin, 88*(3), 588–606.

Bollen, K. A., (1990). Overall fit in covariance structure models: Two types of sample size effects. *Psychological Bulletin, 107*(2), 256–259.

Concannon, F., Flynn, A., & Campbell, (2005). What campus-based students think about the quality and benefits of e-learning. *British Journal of Educational Technology. 36*(2), 501–512.

Davis, F. D., (1986). *A Technology Acceptance Model for Empirically Testing New End-User Information Systems: Theory and Results.* Doctoral dissertation. Cambridge, MA: MIT Sloan School of Management.

Davis, F. D., (1989). Perceived usefulness, perceived ease of use, and user acceptance of information technology. *MIS Quarterly, 13*(3), 319–339.

Economist Intelligence Unit, (2003). *The 2003 E-readiness Rankings.* New York, NY: EIU.

Elloumi, F., (2004). Value chain analysis: A strategic approach to online learning. In: Anderson, A., & Elloumi, F., (eds.), *Theory and Practice of Online Learning* (pp. 61–92). Athabasca, Canada: Athabasca University.

Grandon, E., Alshare, O., & Kwan, O., (2005). Factors influencing student intention to adopt online classes: A cross-cultural study. *Journal of Computing Sciences in Colleges, 20*(4), 46–56.

Hair, J. F., Anderson, R. E., Tatham, R. L., & Black, W. C., (1998). *Multivariate Data Analysis*: (5th edn.) Upper Saddle River, NJ: Prentice Hill.

Hara, N., (2000). Student distress in a web-based distance education course. *Information, Communication and Society, 3*(4), 557–579.

Hoelter, J. W., (1983). The analysis of covariance structures: Goodness-of-fit indices. *Sociological Methods and Research, 11,* 325–344.

Jöreskog, K. G., & Sörbom, D., (1989). *LISREL 7: A Guide to the Programs and Applications*. Chicago, IL: SPSS.

Kelloway, E. K., (1998). *Using LISREL for Structural Equation Modeling: A Researcher's Guide*. Thousand Oaks, CA: Sage.

Kilmurray, J., (2003). *E-learning: It's More Than Automation*. The Technology Source archives. Retrieved from: http://technologysource.org/article/elearning/ (accessed on 09 March 2024).

Kim, C., & Santiago, R., (2005). Constructions of e-learning environments in Korea. *Educational Technology Research and Development, 53*(4), 108–114.

Koohang, A., & Durante, A., (2003). Learners' perceptions toward the web-based distance learning activities/assignments portion of an undergraduate hybrid instructional model. *Journal of Informational Technology Education, 2*, 105–113.

Junghoon Leem, B. L. (2007). The Current Status of e-Learning and Strategies to enhance Educational Competitiveness in Korean Higher Education. International Review of Research in Open and Distance Learning, 1–18.

Legris, P., Ingham, J., & Collerette, P., (2003). Why do people use information technology? A critical review of the technology acceptance model. *Information & Management, 40*, 191–204.

Malhotra, Y., & Galletta, D. F., (1999). Extending the technology acceptance model to account for social influence: Theoretical bases and empirical validation. *Proceedings of the 32nd Hawaii International Conference on System Sciences.*

Marsh, W. H., Balla, J. R., & MacDonald, R. P., (1988). Goodness-of-fit indexes in confirmatory factor analysis: The effect of sample size. *Psychological Bulletin, 88*, 245–258.

CHAPTER 6

Human Resource Disclosure Practices in Select Indian Banks: A Comparative Study of Public and Private Sectors

ARINDAM GHOSH,[1] AMIR JAFAR,[1] ANJAN KUMAR GHOSH,[2] and MAINAK CHAKRABORTY[3]

[1]Associate Professor, Department of Management Science and Humanities, MCKV Institute of Engineering, West Bengal, India

[2]Research Scholar, Department of Business Administration (Human Resource), The University of Burdwan, West Bengal, India

[3]Assistant Professor, Department of Business Management, School of Management, Swami Vivekananda University, West Bengal, India.

ABSTRACT

In this study, the researchers explored the top 10 Indian banks (ranked by revenue earned) from the Fortune India 500 list of 2020. Thorough review of corporate reports and information was conducted to understand the current HR reporting practices. All parameters were assessed based on the disclosed and undisclosed status of HR reports over a period of six years (2015–16 to 2020–21). The findings reveal a clear disparity in HR reporting practices between public and private sector banks. However, regardless of the degree of alignment or mismatch, the majority of reporting practices do not adhere to the objectives of the Companies Act (2013), with a minimal amount of reported HR practices and issues. Additionally, an irregular trend of HR disclosure has been identified. This study has significant implications for banks, other companies, policymakers, and researchers, aiming to develop or enhance HR reporting practices.

Contemporary Digital Transformation and Organizational Effectiveness in Business 4.0.
Nilanjan Ray, Sudin Bag, & Tapas Kumar Chatterjee (Eds.)

6.1 INTRODUCTION

The business practices are constantly evolving in tune with the pace of globalization. The constant paradigm shift compels organizations to adopt new practices in order to survive and thrive. Today's organizations are extremely sensitive to their environment in the pursuit of sustainable competitive advantage. Along with other changes in the business, reporting practices are undergoing a sea change. In India, the enactment of the Companies Act (2013) has introduced a new paradigm for corporate reporting. Global corporate reporting practices also have an impact.

According to Casten, Lynch & Lynch (2014), the global business environment and external pressures such as climate change, resource scarcity, population growth, and worldwide economic development are compelling corporations to focus more on nonfinancial reporting in addition to their financial reporting. There is a constant shift in nonfinancial reporting, from R&D information reporting to CSR and HR reporting (Arvidsson, 2011). Studies reveal that CSR and HR reporting are becoming very popular and an important source of information for effective decision making. Jones & Jonas (2011) remarked that "no longer is it just GAAP or GAAS, there is now CSR and GRI…." Globally, corporations are investing money and time to implement their CSR and HR policies into action, and they are reporting these activities externally for public use. In India, CSR and HR practices and reporting are being set by the state and corporations jointly (Sharma, 2013). During the last decade, especially after the enactment of the Companies Act (2013), these practices have taken shape and are available externally.

In this milieu, it is imperative to know the current nonfinancial reporting practices of Indian companies. In this study, researchers intend to identify the current practices and shifting trends of HR reporting in the Indian banking industry, with a comparative overview of public and private sector banks.

6.2 THEORETICAL BACKGROUND OF HR DISCLOSURE

Human resource (HR) information is widely used by different stakeholders or researchers for effective decision making. Reporting human capital (HC) information or HR information is a part of intellectual capital (IC) disclosure (Vergauwen & Alem, 2005). According to Graaf (2013), as IC is highly dependent on financial indicators, it should be treated sensibly along with financial capital. The author also opined that corporate reporting practices should be improved as it is a very important tool for communication of non-financials to the capital market. Globally, companies are disclosing their IC information in annual reports, financial statements, or IC reports (Vergauwen & Alem, 2005). These reporting practices are generally performed as per the Global Reporting Initiatives (GRI), International Integrated Reporting Council (IIRC), or legal compliance requirements of the respective countries.

The disclosure of HR information can serve various purposes. Research studies have shown that good reporting practices enhance stakeholders' decision-making capabilities, while insufficient or false reporting can be misleading. According to Beattie & Smith (2010), external disclosure is a valuable tool for recruitment and therefore holds great importance for potential job seekers. Vergauwen & Alem (2005) have expressed the view that inadequate IC information can misguide shareholders. Many researchers argue that HR is one of the most crucial factors in corporate value creation. Therefore, voluntary disclosure about HRs also plays a significant role in a firm's value creation, reputation, and image building. This, in turn, enhances shareholders' trust for investment (Dominguez, 2011; Rimmel, Dergård, & Jonäll, 2012; Uyar & Kiliç, 2012). Several studies on nonfinancial reporting have identified noticeable variations in reporting practices (such as level, nature, quality, etc.) based on different factors. HR disclosure in annual reports varies in terms of frequency, content, and the nature of the business (Subbarao & Zeghal, 1997). Lack of knowledge about nonfinancial reporting is one of the major causes of poor reporting practices. The study by Vergauwen & Alem (2005) found that IC reporting varies widely among countries, as well as among companies within the same industry and country (Singh & Kansal, 2011).

Several studies have been conducted to date in the global context, adopting content analysis of annual reports as a method. These studies aimed to identify the content, level, and changes in HR disclosure. The study by Singh & Kansal (2011) found that IC Disclosure practices of listed Indian pharmaceutical companies are low and narrative. Although large companies collect data on employee expenditure for providing different benefits, employee turnover, training costs, workplace safety, etc., they rarely disclose it to the public (Beattie & Smith, 2010; Hayashi, 2003). The study by Branco & Rodrigues (2009) revealed that the 'best companies to work for' disclose more information than the control companies. With the growing demand for HR information, companies should be more transparent about their HR practices (Olsson, 2001). Governments and other regulatory bodies should also be more committed to improving corporate disclosure.

6.3 OBJECTIVES OF THE STUDY

The amended Companies Act of 2013 has mandated the inclusion of HR information in the annual report and/or business responsibility report. Although the Companies Act of 2013 does not apply to banking organizations, banks voluntarily report externally as part of their commitment to good corporate citizenship. Against this backdrop, it is crucial to examine the HR reporting practices of Indian banks. This study aims to explore the extent to which Indian banks have adopted HR reporting in their corporate reports and websites over a six-year period (2015–2016 to 2020–2021). The main objectives of the study are:

- To investigate the current HR disclosure practices of the select Indian banks;
- To understand the changes in the reporting practices of the selected banks over the selected time period; and
- To compare the HR disclosure practices of the selected private and public sector banks.

6.4 METHODOLOGY

This study is based on a comprehensive review of the existing literature on nonfinancial corporate reporting practices from secondary sources. The information regarding HR disclosure has been obtained from the latest available annual reports, business responsibility reports, and corporate websites of selected Indian banks – both from the private and public sectors.

The researchers have considered the top 10 banks in India (by revenue) from the Fortune India 500 list of 2020. The 10 selected Indian banks for this study are *State Bank of India, ICICI Bank, HDFC Bank, Axis Bank, Punjab National Bank, Bank of Baroda, Canara Bank, Bank of India, Kotak Mahindra Bank*, and *Union Bank of India*. Out of these 10 banks, six are public and four are private banks. The corporate reports and information on the websites of the selected companies have been scrutinized to understand the current practices related to HR reporting. All parameters were evaluated based on the disclosed and undisclosed status over a period of six years (2015–2016 to 2020–2021).

HR disclosure is a significant source of nonfinancial information. Numerous studies have identified multiple parameters to assess HR disclosure practices. Alvarez (2011) considered staff performance, compensation, incentives, motivation, interaction with staff, health and safety at work, workers' rights, and industrial relations as general parameters for identifying HR disclosure. Based on previous literature and corporate reports of selected companies, the researchers have identified a set of 14 HR parameters, including employment, training and development, compensation and benefits, employee health and wellbeing, work-life balance, working conditions, employer brand, employee contribution to society, and sexual harassment at workplaces.

The research problem for this study has been formulated based on the aforementioned panel dataset, which consists of 10 selected Indian banks over a period of six years. Due to the non-normal distribution nature, non-parametric tests have been designed with a small sample size of 58 observations.

As mentioned in the previous section, the construct designed for this study based on the literature review is "HR Disclosure Status," which encompasses the aforementioned 14 variables.

The dichotomous scale has been used in this study to document the HR Disclosure Status, and the Chi-Square test of Independence has been performed through cross-tabulation. The independent variable is "Types of Banks" with two categories – public and private. The dependent variable is "HR Disclosure Status."

The hypothesis for this study is that "the practice of HR disclosure is not dependent upon the types of Indian banks – public or private" (H_0).

However, this study has tested the relations of the independent variable with all the variables under the construct of "HR Disclosure Status" separately through the Chi-Square test.

6.4.1 MAJOR OBSERVATIONS OF THE STUDY

A comparative presentation of HR disclosure practices of private and public sector banks has been shown in the comparative bar chart (Figure 6.1), which illustrates the remarkable difference in reporting practices in a few cases, while being very close to each other in most instances (data expressed as a percentage of total reporting provisions).

Similarly, the trend line reflecting this comparison is shown in Figure 6.2. This figure displays the median of the total reporting made by both sectors. The fluctuations, along with the recent downtrend, are noteworthy in this diagram.

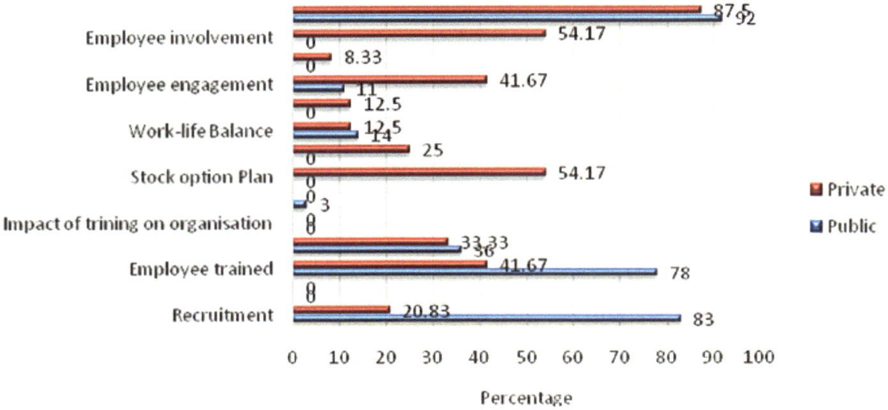

FIGURE 6.1 Comparative status of HR disclosure (in percentage).

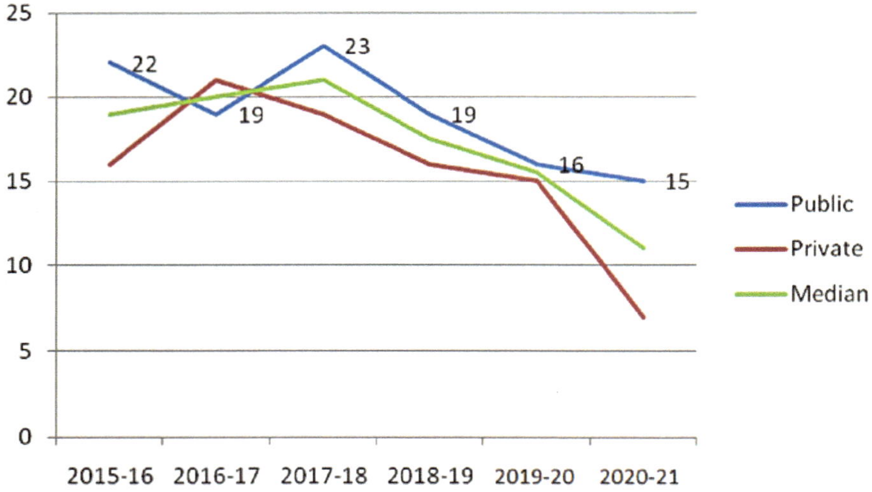

FIGURE 6.2 Comparative status of HR disclosure trends (in total count).

From this study, it has been observed that HR reporting practices are still a neglected area in annual reports. Only a few pieces of information, such as details of employees (e.g., total number of women in the workforce, etc.), training details, and sexual harassment, are available as some parts of these information disclosures are mandatory for the select bank. In some cases, a few other HR-related information is available. HR information is mostly qualitative in nature and too limited to gain an idea or make decisions.

Based on the content analysis of the annual report and corporate websites, it can be said that these top Indian banks are reluctant to disclose their HR information and sinking changes have been observed over a period of time. Banks should take appropriate measures to improve HR disclosure. The government and other regulatory bodies may take initiatives and impose certain rules and regulations.

6.5 ANALYSIS OF DATA

The relationship between the independent variable, Types of Banks, and the set of dependent variables under the construct – HR disclosure practice, has been examined in this study using the Chi-Square test of Independence. The significance of these relationships is presented in subsections.

6.5.1 RELATION BETWEEN TYPES OF BANKS AND SEXUAL HARASSMENT REPORTING

The hypothesis for this study is that "the practice of Sexual Harassment Reporting is not dependent upon the types of Indian banks – public or private" (H_0).

The cross-tabulation in Table 6.1 reveals that most of the public banks (33 out of 34) have reported the sexual harassment status of their organization in their nonfinancial reports over a period of six years (2015–2016 to 2020–2021) since the enforcement of the Companies Act (2013). The same applies to private banks, with 21 out of 24 reporting provisions.

Both the tests of asymptomatic significance (Pearson Chi-Square) and exact significance (Fisher's Exact Test) have shown no significance in this Chi-Square test. Therefore, it can be concluded that the sexual harassment reporting practice does not vary according to the types of banks and remains the same for private and public banks in India.

6.5.2 RELATION BETWEEN TYPES OF BANKS AND EMPLOYEE INVOLVEMENT REPORTING

In this study, it has been hypothesized that "the practice of Employee Involvement Reporting is not dependent upon the types of Indian banks – public or private" (H_0).

Table 6.2 shows that none of the public banks have reported the employee involvement status [0 out of 34] in their nonfinancial reports from 2015–2016

TABLE 6.1 Cross Tabulation with Chi-Square Test Results: Sexual Harassment Reporting

		Sexual Harassment Reporting	
		No	**Yes**
Types of Banks	Public banks	1	33
	Private banks	3	21
Total		4	54

	Value	df	Asymptotic Significance (Two-Sided)	Exact Sig. (Two-Sided)	Exact Sig. (One-Sided)
Pearson Chi-Square	2.002[a]	1	0.157	–	–
Continuity Correction[b]	0.790	1	0.374	–	–
Likelihood Ratio	2.003	1	0.157	–	–
Fisher's Exact Test	–	–	–	0.297	0.187
Linear-by-Linear Association	1.968	1	0.161	–	–
Number of Valid Cases	58	–	–	–	–

[a] Two cells (50.0%) have expected count less than 5. The minimum expected count is 1.66.

[b] Computed only for 2x2 table.

to 2020–2021. However, the private banks have 13 reports out of 24 reporting provisions. The chi-square test results are shown in Table 6.2.

TABLE 6.2 Cross Tabulation with Chi-Square Test Results: Employee Involvement Reporting

		Employee Involvement Reporting	
		No	**Yes**
Types of Banks	Public banks	34	0
	Private banks	11	13
Total		45	13

	Value	df	Asymptotic Significance (Two-Sided)	Exact Sig. (Two-Sided)	Exact Sig. (One-Sided)
Pearson Chi-Square	23.737[a]	1	0.000	–	–
Continuity Correction[b]	20.724	1	0.000	–	–
Likelihood Ratio	28.619	1	0.000	–	–
Fisher's Exact Test	–	–	–	0.000	0.000
Linear-by-Linear Association	23.328	1	0.000	–	–
Number of Valid Cases	58	–	–	–	–

[a] 0 cells (0.0%) have expected count less than 5. The minimum expected count is 5.38.

[b] Computed only for 2x2 table.

Both the tests of asymptotic significance and exact significance have shown significance in this Chi-Square test. Therefore, it can be concluded that the employee involvement reporting practices vary significantly according to the types of banks in India.

6.5.3 RELATION BETWEEN TYPES OF BANKS AND EMPLOYER BRANDING REPORTING

The hypothesis formulated for this study is that "the practice of Employer Branding Reporting is not dependent upon the types of Indian banks – public or private" (H_0).

The cross-tabulation (Table 6.3) shows that none of the public banks have reported the employer branding status [0 out of 34] of their organization in their nonfinancial reports over the span of six years (2015–2016 to 2020–2021). However, the private banks have 2 reports out of 24 reporting provisions.

TABLE 6.3 Cross Tabulation with Chi-Square Test Results: Employer Branding Reporting

		Employer Branding Reporting	
		No	**Yes**
Types of Banks	Public banks	34	0
	Private banks	22	2
Total		56	2

	Value	**df**	**Asymptotic Significance (Two-Sided)**	**Exact Sig. (Two-Sided)**	**Exact Sig. (One-Sided)**
Pearson Chi-Square	2.935[a]	1	0.087	–	–
Continuity Correctionb	0.965	1	0.326	–	–
Likelihood Ratio	3.631	1	0.057	–	–
Fisher's Exact Test	–	–	–	0.167	0.167
Linear-by-Linear Association	2.884	1	0.089	–	–
Number of Valid Cases	58	–	–	–	–

[a] Two cells (50.0%) have expected count less than 5. The minimum expected count is 0.83.

[b] Computed only for 2x2 table.

Both the tests of asymptomatic significance and exact significance have not been significant in this Chi-Square test. Therefore, it can be stated the employer branding reporting practice does not vary significantly according to the types of banks in India.

6.5.4 RELATION BETWEEN TYPES OF BANKS AND EMPLOYEE ENGAGEMENT REPORTING

The hypothesis formulated for this study is "the practice of Employee Engagement Reporting is not contingent upon the categories of Indian banks – public or private" (H_0).

The subsequent Table 6.4 demonstrates a limited number of public banks have documented the employee engagement status [4 out of 34] of their organization in their nonfinancial reports over the course of six years (2015–2016 to 2020–2021). However, the private banks have 10 reports out of 24 reporting provisions.

TABLE 6.4 Cross Tabulation with Chi-Square Test Results: Employee Engagement Reporting

		Employee Engagement Reporting	
		No	**Yes**
Types of Banks	Public banks	30	4
	Private banks	14	10
Total		44	14

	Value	df	Asymptotic Significance (Two-Sided)	Exact Sig. (Two-Sided)	Exact Sig. (One-Sided)
Pearson Chi-Square	6.870[a]	1	0.009	–	–
Continuity Correction[b]	5.334	1	0.021	–	–
Likelihood Ratio	6.878	1	0.009	–	–
Fisher's Exact Test	–	–	–	0.013	0.011
Linear-by-Linear Association	6.751	1	0.009	–	–
Number of Valid Cases	58	–	–	–	–

[a] 0 cells (.0%) have expected count less than 5. The minimum expected count is 5.79.

[b] Computed only for 2x2 table.

Both the examinations of asymptotic significance and precise significance have yielded statistically significant results in this Chi-Square test. Therefore, it can be asserted that the reporting practices of employee engagement vary significantly based on the types of banks in India.

6.5.5 RELATION BETWEEN TYPES OF BANKS AND WORK CULTURE REPORTING

We hypothesize for this study that the practice of Work Culture Reporting is not dependent upon the types of Indian banks – public or private (H_0).

The cross-tabulation in Table 6.5 shows that none of the public banks have reported the work culture status [0 out of 34] of their organization in their nonfinancial reports over the span of six years (2015–2016 to 2020–2021). However, 3 private banks have reported the work culture out of 24 reporting provisions.

TABLE 6.5 Cross Tabulation with Chi-Square Test Results: Work-Culture Reporting

		Work Culture Reporting	
		No	**Yes**
Types of Banks	Public banks	34	0
	Private banks	21	3
Total		55	3

	Value	df	Asymptotic Significance (Two-Sided)	Exact Sig. (Two-Sided)	Exact Sig. (One-Sided)
Pearson Chi-Square	4.482[a]	1	0.034	–	–
Continuity Correction[b]	2.296	1	0.130	–	–
Likelihood Ratio	5.528	1	0.019	–	–
Fisher's Exact Test	–	–	–	0.066	0.066
Linear-by-Linear Association	4.405	1	0.036	–	–
Number of Valid Cases	58	–	–	–	–

[a] Two cells (50.0%) have expected count less than 5. The minimum expected count is 1.24.

[b] Computed only for 2x2 table.

It is noteworthy that the asymptomatic distribution is significant with $p<0.05$. Therefore, based on the Chi-Squared approximation, it can be interpreted that the null hypothesis is not accepted. However, considering the small sample size in this case, the Fisher's Exact Test (Exact significance) has been taken into account, which is not significant with $p>0.05$. Therefore, it can be concluded that the work culture reporting practice does not vary significantly according to the types of banks in India.

6.5.6 RELATION BETWEEN TYPES OF BANKS AND WORK-LIFE BALANCE REPORTING

The hypothesis for this study is that the practice of Work Life Balance Reporting is not dependent upon the types of Indian banks – public or private (H_0).

Table 6.6 shows that a few public banks have reported the work-life balance status [4 out of 34] of their organization in their nonfinancial reports over six years (2015–2016 to 2020–2021). Similarly, 3 private banks have reported the work-life balance out of 24 reporting provisions.

TABLE 6.6 Cross Tabulation with Chi-Square Test Results: Work-Life Balance Reporting

		Work Life Balance Reporting	
		No	**Yes**
Types of Banks	Public banks	30	4
	Private banks	21	3
Total		51	7

	Value	df	Asymptotic Significance (Two-Sided)	Exact Sig. (Two-Sided)	Exact Sig. (One-Sided)
Pearson Chi-Square	0.007[a]	1	0.933	–	–
Continuity Correctionb	0.000	1	1.000	–	–
Likelihood Ratio	0.007	1	0.933	–	–
Fisher's Exact Test	–	–	–	1.000	0.619
Linear-by-Linear Association	0.007	1	0.933	–	–
Number of Valid Cases	58	–	–	–	–

[a] Two cells (50.0%) have expected count less than 5. The minimum expected count is 2.90.

[b] Computed only for 2x2 table.

Both the tests for asymptomatic significance and exact significance were not significant in this Chi-Square test. Therefore, it can be stated the work-life balance reporting practice does not vary significantly according to the types of banks in India.

6.5.7 RELATION BETWEEN TYPES OF BANKS AND EMPLOYEE HEALTH AND WELL-BEING REPORTING

We have formulated the hypothesis for this study as follows: "The practice of Employee Health and Well Being Reporting is not dependent upon the types of Indian banks – public or private" (H_0).

The cross-tabulation presented in Table 6.7 reveals that none of the public banks have reported the employee health and well-being status [0 out of 34] of their organization in their nonfinancial reports over the span of six years (2015–2016 to 2020–2021). However, 6 private banks have included the employee health and well-being in their reporting, out of the 24 reporting provisions.

TABLE 6.7 Cross Tabulation with Chi-Square Test Results: Employee Health and Well-Being Reporting

		Employee Health and Well-Being Reporting	
		No	**Yes**
Types of Banks	Public banks	34	0
	Private banks	18	6
Total		52	6

	Value	df	Asymptotic Significance (Two-Sided)	Exact Sig. (Two-Sided)	Exact Sig. (One-Sided)
Pearson Chi-Square	9.481[a]	1	0.002	–	–
Continuity Correction[b]	6.977	1	0.008	–	–
Likelihood Ratio	11.589	1	0.001	–	–
Fisher's Exact Test	–	–	–	0.003	0.003
Linear-by-Linear Association	9.317	1	0.002	–	–
Number of Valid Cases	58	–	–	–	–

[a] Two cells (50.0%) have expected count less than 5. The minimum expected count is 2.48.

[b] Computed only for 2x2 table.

Both the tests of asymptotic significance and exact significance have yielded statistically significant results in this Chi-Square test. Therefore, it can be asserted that the reporting practice of Employee Health and Well Being varies significantly based on the types of banks in India.

6.5.8 REGARDING THE RELATIONSHIP BETWEEN TYPES OF BANKS AND STOCK OPTION PLAN REPORTING

The hypothesis formulated for this study is that "the practice of Stock Option Plan Reporting is not contingent upon the types of Indian banks – public or private" (H_0).

According to Table 6.8, none of the public banks have reported the stock option plan status [0 out of 34] of their organization in their nonfinancial reports over the span of six years (2015–2016 to 2020–2021). However, 13 private banks have reported the stock option plan status out of 24 reporting provisions.

TABLE 6.8 Cross Tabulation with Chi-Square Test Results: Stock Option Plan Reporting

		Stock Option Plan Reporting	
		No	**Yes**
Types of Banks	Public banks	34	0
	Private banks	11	13
Total		45	13

	Value	df	Asymptotic Significance (Two-Sided)	Exact Sig. (Two-Sided)	Exact Sig. (One-Sided)
Pearson Chi-Square	23.737[a]	1	0.000	–	–
Continuity Correction[b]	20.724	1	0.000	–	–
Likelihood Ratio	28.619	1	0.000	–	–
Fisher's Exact Test	–	–	–	0.000	0.000
Linear-by-Linear Association	23.328	1	0.000	–	–
Number of Valid Cases	58	–	–	–	–

[a] 0 cells (0.0%) have expected count less than 5. The minimum expected count is 5.38.

[b] Computed only for 2x2 table.

Both the tests of AS and ES have been significant in this Chi-Square test. So, it can be said that the SOP reporting practice varies significantly according to the types of banks in India.

6.5.9 RELATION BETWEEN TYPES OF BANKS AND INCENTIVE SCHEME REPORTING

The hypothesis is that the practice of Incentive Scheme Reporting is not dependent upon the types of Indian banks – public or private (H_0).

The data in Table 6.9 shows that none of the public banks have reported the incentive schemes [1 out of 34] of their organization in their nonfinancial reports over the span of six years (2015–2016 to 2020–2021). However, no private banks have reported the incentive schemes out of 24 reporting provisions.

TABLE 6.9 Cross Tabulation with Chi-Square Test Results: Incentive Scheme Reporting

		Incentive Scheme Reporting	
		No	Yes
Types of Banks	Public banks	33	1
	Private banks	24	0
Total		57	1

	Value	df	Asymptotic Significance (Two-Sided)	Exact Sig. (Two-Sided)	Exact Sig. (One-Sided)
Pearson Chi-Square	0.718[a]	1	0.397	–	–
Continuity Correction[b]	0.000	1	1.000	–	–
Likelihood Ratio	1.081	1	0.299	–	–
Fisher's Exact Test	–	–	–	1.000	0.586
Linear-by-Linear Association	0.706	1	0.401	–	–
Number of Valid Cases	58	–	–	–	–

[a] Two cells (50.0%) have expected count less than 5. The minimum expected count is 0.41.

[b] Computed only for 2x2 table.

Both the tests of asymptotic significance and exact significance were not significant in this Chi-Square test. Therefore, it can be concluded that the reporting practice of the incentive scheme does not vary significantly according to the types of banks in India.

Regarding the relationship between the types of banks and the impact of training on organization reporting, we formulated the hypothesis for this study as follows:

"The practice of Impact of Training on Organization Reporting is not dependent upon the types of Indian banks – public or private" (H_0).

The cross-tabulation presented in Table 6.10 reveals that none of the public and private banks reported the impact of training on organization (0 out of 34) in their nonfinancial reports over the span of six years (2015–2016 to 2020–2021).

TABLE 6.10 Cross Tabulation with Chi-Square Test Results: Impact of Training on Organization Reporting

		Impact of Training on Organization Reporting
		No
Types of Banks	Public banks	34
	Private banks	24
Total		58

In this case, no stats are calculated because the Impact of Training on Organization reporting is always the same, so it is not considered as a variable here.

6.5.10 RELATION BETWEEN TYPES OF BANKS AND TIME SPENT FOR TRAINING REPORTING

The hypothesis for this study is as follows: "the practice of Time Spent for Training Reporting is not dependent upon the types of Indian banks – public or private" (H_0).

Table 6.11 shows that 13 public banks have reported the time spent for training reporting status [13 out of 34] of their organization in their nonfinancial reports over the span of six years (2015–2016 to 2020–2021). Around 8 private banks have reported the time spent for training reporting out of 24 reporting provisions.

TABLE 6.11 Cross Tabulation with Chi-Square Test Results: Time Spent for Training Reporting

		Time Spent for Training Reporting	
		No	**Yes**
Types of Banks	Public banks	21	13
	Private banks	16	8
Total		37	21

	Value	Df	Asymptotic Significance (Two-Sided)	Exact Sig. (Two-Sided)	Exact Sig. (One-Sided)
Pearson Chi-Square	0.146[a]	1	0.702	–	–
Continuity Correction[b]	0.011	1	0.916	–	–
Likelihood Ratio	0.147	1	0.701	–	–
Fisher's Exact Test	–	–	–	0.786	0.460
Linear-by-Linear Association	0.144	1	0.704	–	–
Number of Valid Cases	58	–	–	–	–

[a] 0 cells (0.0%) have expected count less than 5. The minimum expected count is 8.69.

[b] Computed only for 2x2 table.

The Chi-Square test results indicate that neither the tests of asymptotic significance nor exact significance show any statistical significance.

Therefore, we can conclude that the time spent on training reporting practice does not significantly differ based on the types of banks in India.

6.5.11 THE RELATIONSHIP BETWEEN TYPES OF BANKS AND REPORTING OF EMPLOYEES TRAINED

In this study, the hypothesis is that "the practice of Employees Trained Reporting is not influenced by the types of Indian banks – public or private" (H_0).

According to Table 6.12, out of 34 public banks, 26 have reported the employee trained status of their organization in their nonfinancial reports over a span of six years (2015–2016 to 2020–2021). Among the 24 reporting provisions, 10 private banks have reported the employee trained status.

TABLE 6.12 Cross Tabulation with Chi-Square Test Results: Employees Trained Reporting

		Employees Trained Reporting	
		No	**Yes**
Types of Banks	Public banks	8	26
	Private banks	14	10
Total		22	36

	Value	df	Asymptotic Significance (Two-Sided)	Exact Sig. (Two-Sided)	Exact Sig. (One-Sided)
Pearson Chi-Square	7.239[a]	1	0.007	–	–
Continuity Correction[b]	5.836	1	0.016	–	–
Likelihood Ratio	7.290	1	0.007	–	–
Fisher's Exact Test	–	–	–	0.013	0.008
Linear-by-Linear Association	7.114	1	0.008	–	–
Number of Valid Cases	58	–	–	–	–

[a] 0 cells (0.0%) have expected count less than 5. The minimum expected count is 9.10.

[b] Computed only for 2x2 table.

Both the tests for asymptotic significance and exact significance have produced statistically significant outcomes in this Chi-Square test. Hence, we can infer that the reporting practice of Employees Trained varies significantly based on the types of banks in India.

6.5.12 REGARDING THE RELATIONSHIP BETWEEN TYPES OF BANKS AND ABSENTEEISM REPORTING

For this study, we have formulated the hypothesis that "the practice of Absenteeism Reporting is not contingent upon the types of Indian banks – public or private" (H_0).

The provided cross-tabulation (Table 6.13) reveals that neither public nor private banks have included the employee absenteeism status [0 out of 34] of their organization in their nonfinancial reports over the course of six years (2015–2016 to 2020–2021).

TABLE 6.13 Cross Tabulation with Chi-Square Test Results: Absenteeism Reporting

		Absenteeism Reporting
		No
Types of Banks	Public banks	34
	Private banks	24
Total		58

In this case, no statistics are computed because Absenteeism reporting is a constant and therefore should not be considered as a variable.

6.5.13 RELATION BETWEEN TYPES OF BANKS AND RECRUITMENT REPORTING

For this study, the proposition is that "the practice of Recruitment Reporting is not contingent upon the categories of Indian banks – public or private" (H_0).

The cross-tabulation provided in Table 6.14 demonstrates that 29 public banks had reported the recruitment status [26 out of 34] of their organization in their nonfinancial reports over the course of six years (2015–2016 to 2020–2021). Conversely, 5 private banks had reported the recruitment status out of 24 reporting provisions.

TABLE 6.14 Cross Tabulation with Chi-Square Test Results: Recruitment Reporting

		Recruitment Reporting	
		No	**Yes**
Types of Banks	Public banks	5	29
	Private banks	19	5
Total		24	34

TABLE 6.14 *(Continued)*

	Value	df	Asymptotic Significance (Two-Sided)	Exact Sig. (Two-Sided)	Exact Sig. (One-Sided)
Pearson Chi-Square	24.100[a]	1	0.000	–	–
Continuity Correction[b]	21.516	1	0.000	–	–
Likelihood Ratio	25.714	1	0.000	–	–
Fisher's Exact Test	–	–	–	0.000	0.000
Linear-by-Linear Association	23.685	1	0.000	–	–
Number of Valid Cases	58	–	–	–	–

[a] 0 cells (0.0%) have expected count less than 5. The minimum expected count is 9.93.

[b] Computed only for 2x2 table.

Both the tests of asymptotic significance and exact significance have yielded significant results in this Chi-Square test. Therefore, it can be concluded that the reporting practices for recruitment vary significantly according to the types of banks in India.

6.6 FINDINGS

From the above analysis, it is observed that there have been some significant differences in HR disclosure practices between private and public sector banks in India. The following reporting areas are found to be different in private and public sector banks in terms of reporting practices:

- Employee involvement;
- Employee engagement;
- Employee health and well-being;
- Stock option plan;
- Employees trained;
- Recruitment.

Out of the 14 HR disclosure parameters, only 8 parameters have been consistently adhered to for disclosing HR practices in Indian banks, both private and public. However, this consistency has a negative implication as private and public sector banks have limited reporting on HR issues. The only area where both sectors have a significant amount of HR reports is Sexual Harassment.

6.7 CONCLUSION, LIMITATIONS, AND SCOPE FOR FURTHER RESEARCH

Although nonfinancial reporting practices have gained momentum in developed countries, in India, it takes years to achieve full-fledged practices in corporate reporting. It can be said that the legalization of mandatory disclosure of business responsibility reports by top Indian companies will definitely boost nonfinancial reporting practices. Companies should disclose a single integrated report containing information related to their intellectual capital (IC), social and environmental accountability, as it is always beneficial for stakeholders (Oliveira, Rodrigues, & Craig, 2010). It has also been observed that HR reporting practices are still very poor and have shown a continuous negative change over the years. It has been found that very few HR items have been disclosed, and the majority of them were part of mandatory disclosure. Despite being the most important resources of companies, external reporting of HR is still neglected. Companies and states may formulate policies regarding adequate HR reporting for improved corporate reports and better decision-making. This study will definitely help carry out further research on nonfinancial reporting in industrial segments other than banks. This study may also assist companies and states in formulating nonfinancial reporting policies.

This study is based on the top 10 banks operating in India. The study could be more comprehensive if all banks were included. Additionally, there is no benchmark in HR reporting for the banking industry, so the degree of information disclosed could not be determined in this study. The information that was disclosed or not disclosed was based on the objective findings obtained from the reports. Further studies may be conducted to develop a nonfinancial disclosure matrix or a model of HR reporting for the banking industry.

KEYWORDS

- **banking industry**
- **employee involvement**
- **HR disclosure**
- **Indian banking sector**
- **nonfinancial disclosure matrix**
- **nonfinancial reporting**
- **public sector banks**
- **stakeholders**

REFERENCES

Alvarez, A., (2015). Corporate response to human resource disclosure recommendations. *Social Responsibility Journal, 11*(2), 306–323. https://doi.org/10.1108/SRJ-06-2013-0076.

Arvidsson, S., (2011). Disclosure of non-financial information in the annual report: A management-team perspective. *Journal of Intellectual Capital, 12*(2), 6. https://doi.org/10.1108/ 14691931111123421.

Ballou, B., Heitger, D. L., & Landes, C. E., (2006). The future of corporate sustainability reporting. *Journal of Accountancy, 202*(90), 65–74.

Beardsell, J., (2008). *The Influence of CSR Disclosure on Corporate Governance and Company Performance.* SMC Working Paper, 20. https://doi.org/10.2139/ssrn.1302314.

Beattie, V., & Smith, S. J., (2010). Human capital, value creation and disclosure. *Journal of Human Resource Costing & Accounting, 14*(4), 262–285. https://doi.org/10.1108/14013381011105957.

Branco, M., & Rodrigues, L., (2009). Exploring the importance of social responsibility disclosure for human resources. *Journal of Human Resource Costing & Accounting, 13*(3), 186–205. https://doi.org/10.1108/14013380910995494.

Cahan, S. F., De Villiers, C., Jeter, D. C., Naiker, V., & Van, S. C. J., (2015). Are CSR disclosures value relevant? Cross-country evidence. *European Accounting Review, 8180,* 1–33. https://doi.org/10.1080/09638180.2015.1064009.

Casten, D. B., Lynch, N. C., & Lynch, M. F., (2014). The expanding use of sustainability reporting. Standards setting and assurance opportunities for CPAs. *The CPA Journal, 84*(3), 18–25.

Charles, C. H., Giovanna, M., Dennis, P. M., & Robin, W. R., (2015). CSR disclosure: The more things change…? *Accounting, Auditing & Accountability Journal, 28*(1), 14–35. https://doi.org/10.1108/AAAJ-12-2013-1549.

Cohen, J. R., & Simnett, R., (2015). *CSR and Assurance Services: A Research Agenda, 34*(1), 59–74. https://doi.org/10.2308/ajpt-50876.

Cohen, J. R., Holder-Webb, L. L., Nath, L., & Wood, D., (2012). Corporate reporting of nonfinancial leading indicators of economic performance and sustainability. *Accounting Horizons, 26*(1), 65–90. https://doi.org/10.2308/acch-50073.

Cohen, J. R., Holder-Webb, L., & Zamora, V. L., (2015). Nonfinancial information preferences of professional investors. *Behavioral Research in Accounting, 27*(2), 127–153. https://doi.org/10.2308/bria-51185.

Dilling, P. F. A., (2010). Sustainability reporting in a global context: What are the characteristics of corporations that provide high quality sustainability reports – an empirical analysis. *The International Business & Economics Research Journal, 9*(1), 19–30.

Dominguez, A. A., (2011). The impact of human resource disclosure on corporate image. *Journal of Human Resource Costing & Accounting, 15*(4), 279–298. https://doi.org/10.1108/14013381111197225.

Dumay, J., & Cai, L., (2015). Using content analysis as a research methodology for investigating intellectual capital disclosure: A critique. *Journal of Intellectual Capital, 16*(1), 121–155. https://doi.org/10.1108/JIC-04-2014-0043.

Ellen, R., (2009). Just good business: CSR reporting where to start? *New Zealand Management, 56*(1).

Ghosh, S., (2015). Proactive communication of CSR in India: A distant dream or reality? *Paradigm, 19*(2), 115–136. https://doi.org/10.1177/0971890715609845.

Graaf, J., (2013). Coloring the numbers – on the role of intellectual capital in financial reporting. *Journal of Intellectual Capital, 14*(3), 376–394. https://doi.org/10.1108/JIC-03-2013-0037.

Guthrie, J., & Abeysekera, I., (2006). Content analysis of social, environmental reporting: What is new? *Journal of Human Resource Costing & Accounting, 10*(2), 114–126. https://doi.org/10.1108/14013380610703120.

Hayashi, A. M., (2003). HR information disclosure. *MIT Sloan Management Review, 44*(3), 12.

Jafar, A., & Ghosh, A. K., (2015). Human resource disclosure practices and corporate social responsibility reporting in select Indian companies. *Journal of Accounting and Finance, 29*(2), 26–36.

Jones, A., & Jonas, G. A., (2011). Corporate social responsibility reporting: The growing need for input from the accounting profession. *The CPA Journal, 81*(2), 65–71.

Kamath, B., (2008). Intellectual capital disclosure in India: Content analysis of "TecK" firms. *Journal of Human Resource Costing & Accounting, 12*(3), 213–224. https://doi.org/10.1108/14013380810919859.

Kavitha, W., & Anita, P., (2011). Disclosures about CSR practices: A literature review. *IUP Journal of Corporate Governance, X*(1), 45–56.

KPMG, (2015). *The KPMG Survey of Corporate Responsibility Reporting 2015.*

Litalien, B. C., & Miller, J., (2012). Corporate social responsibility and generation Y: A critical convergence for franchising. *Franchising World, 44*(8), 43–44.

Low, M., Samkin, G., & Li, Y., (2015). Voluntary reporting of intellectual capital: Comparing the quality of disclosures from New Zealand, Australian and United Kingdom universities. *Journal of Intellectual Capital, 16*(4), 779–808. https://doi.org/10.1108/JIC-03-2015-0022.

Mögele, B., & Tropp, J., (2010). The emergence of CSR as an advertising topic: A longitudinal study of German CSR advertisements. *Journal of Marketing Communications, 16*(3), 163–181. https://doi.org/10.1080/13527260802648359.

Naiker, V., (2013). Discussion of "strange bedfellows? Voluntary CSR disclosure and politics." *Accounting and Finance, 53*(4), 905–911. https://doi.org/10.1111/acfi.12052.

Narayan, P. S., Lal, N., Dutta, A., Mehta, T., Majumdar, A., Madhavan, A., & Poor, U., (2012). Corporate social responsibility: Practice, theory, and challenges. *Vikalpa, 37*(2), 73–77.

Narwal, M., (2007). CSR initiatives of indian banking industry. *Social Responsibility Journal, 3*(4), 49–60. https://doi.org/10.1108/17471110710840233.

Oliveira, L., Rodrigues, L. L., & Craig, R., (2010). Intellectual capital reporting in sustainability reports. *Journal of Intellectual Capital, 11*(4), 575–594. https://doi.org/10.1108/ 14691931011085696.

Olsson, B., (2001). Annual reporting practices: Information about human resources in corporate annual reports in major Swedish companies. *Journal of Human Resource Costing & Accounting, 6*(1), 39–52. https://doi.org/10.1108/eb029071.

Rimmel, G., Dergård, J., & Jonäll, K., (2012). Human resources disclosure in Danish intellectual capital statements. *Journal of Human Resource Costing & Accounting, 16*(2), 112–141. https://doi.org/10.1108/14013381211284245.

Roberts, R. W., (n.d.). *Analyses Company Practices in Disclosing Information About the Environmental and Social Impacts of Their Activities*, 83–85.

Schneider, A., & Samkin, G., (2008). Intellectual capital reporting by the New Zealand local government sector. *Journal of Intellectual Capital* (Vol. 9). https://doi.org/10.1108/14691930810892036.

Sharma, S., (2013). Corporate social responsibility in India – the emerging discourse & concerns. *The Indian Journal of Industrial Relations*, *48*(4), 582–597.

Shroff, S., Pilot, S., Kapoor, R., & Bagchi, S., (2013). FOCUS on record/CSR clause in new companies bill. *Business Today*, *22*(19), 20.

Singh, P., & Sarkar, S., (2013). CSR guidelines for Indian companies. *Indian Journal of Corporate Governance*, *6*(1), 76–89.

Singh, S., & Kansal, M., (2011). Voluntary disclosures of intellectual capital. *Journal of Intellectual Capital*, *12*(2), 301–318. https://doi.org/10.1108/14691931111123430.

Srivastava, R. P., Rao, S. S., & Mock, T. J., (2013). Planning and evaluation of assurance services for sustainability reporting: An evidential reasoning approach. *Journal of Information Systems*, *27*(2), 107–126. https://doi.org/10.2308/isys-50564.

Subbarao, A. V., & Zeghal, D., (1997). Human resources information disclosure in annual reports: An international comparison. *Journal of Human Resource Costing & Accounting*, *2*(2), 53–73. https://doi.org/10.1108/eb029039.

Thorne, L., Mahoney, L. S., & Manetti, G., (2014). Motivations for issuing standalone CSR reports: A survey of Canadian firms. *Accounting, Auditing & Accountability Journal* (Vol. 27). https://doi.org/10.1108/AAAJ-07-2013-1393.

Uyar, A., & Kiliç, M., (2012). Value relevance of voluntary disclosure: Evidence from Turkish firms. *Journal of Intellectual Capital*, *13*(3), 363–376. https://doi.org/10.1108/14691931211248918.

Vergauwen, P. G. M. C., & Alem, F. J. C. V., (2005). Annual report IC disclosures in the Netherlands, France, and Germany. *Journal of Intellectual Capital*, *6*(1), 89–104. https://doi.org/10.1108/14691930510574681.

Waller, D. S., & Lanis, R., (2009). Corporate social responsibility (CSR) disclosure of advertising agencies: An exploratory analysis of six holding companies' annual reports. *Journal of Advertising*, *38*(1), 109–122. https://doi.org/10.2753/JOA0091-3367380107.

Withisuphakorn, P., & Jiraporn, P., (2015). The effect of firm maturity on corporate social responsibility (CSR): Do older firms invest more in CSR? *Applied Economics Letters*, *4851*, 1–4. https://doi.org/10.1080/13504851.2015.1071464.

Wong, R., & Millington, A., (2014). Corporate social disclosures: A user perspective on assurance. *Accounting, Auditing & Accountability Journal*, *27*(5), 863–887. https://doi.org/10.1108/AAAJ-06-2013-1389.

Zahller, K. A., Arnold, V., & Roberts, R. W., (2015). Using CSR disclosure quality to develop social resilience to exogenous shocks: A test f investor perceptions. *Behavioral Research in Accounting*, *27*(2), 155–177. https://doi.org/10.2308/bria-51118.

CHAPTER 7

An Investigation into How Artificial Intelligence (AI) has Enhanced Quality Engineering

RACHANA JAISWAL

Assistant Professor, Department of Business Management, HNB Garhwal (A Central) University, Srinagar, Uttarakhand, India

ABSTRACT

In recent years, artificial intelligence (AI) and machine learning (ML) have evolved from cutting-edge practices in various industries to becoming mainstream practices. These advancements have opened up a wide range of applications, including facial recognition and autonomous vehicles. AI stands for artificial intelligence, while ML is an abbreviation for ML (Lenz et al., 2013). Recent years have witnessed significant upheaval, albeit not comparable to the recent disruptions in the field of software development. Businesses are now compelled to adapt at an accelerated pace due to the growing consumer demand for enhanced connectivity through digital experiences. This demand is a direct result of the rise of digitalization. It marks a significant departure from past practices. Attracting and retaining clients by providing comprehensive products and services that meet their needs is now considered "business critical." This factor plays a crucial role in determining whether customers choose to do business with you or your competitors, as well as whether they remain loyal. While the field of Quality Engineering has largely remained unchanged in recent years, every aspect of how quality is delivered has experienced significant expansion. In today's digital-first world, companies are expected to excel in both quality and

Contemporary Digital Transformation and Organizational Effectiveness in Business 4.0.
Nilanjan Ray, Sudin Bag, & Tapas Kumar Chatterjee (Eds.)

speed, eliminating the need to choose between the two. The era of selecting one of the two available options is no longer feasible.

7.1 INTRODUCTION

Businesses are currently engaged in a race on the intensely competitive market to satisfy the increasing needs of their customers and maximize the accessibility of their goods and services. In other words, businesses are striving to meet customer needs while ensuring their goods and services are easily accessible to the intended consumers (Harman, 2012). The primary objective of this competition is to increase the sales of companies' goods and services to their target consumers (Budman et al., 2020). Companies are putting significant effort into ensuring the quality of their goods and services surpasses customer expectations, as these customers are the ultimate purchasers. To maintain a competitive advantage, businesses have adopted agile practices and digital technology. As a result, they have shifted from conventional testing methods to platform-led quality engineering to keep up with the pace of competition (Lambert, 2021).

At the turn of the century, companies from around the world started collaborating on the establishment of independent verification and validation teams. During that time, testing was primarily a centralized approach, separate from the development team. Towards the end of the first decade of the 2000s, a significant number of large and medium-sized businesses invested in creating testing centers of excellence to standardize various methods, tools, and procedures (Li et al., 2020). These investments aimed to enhance the quality of testing and were made by companies from different countries worldwide. The independent approach proved successful for the majority of organizations that adopted the waterfall methodology, contributing to their overall success. However, with the growing popularity of agile methods, customers became more aware of the challenges associated with the autonomous model (Black et al., 2022).

7.2 EVOLUTION OF QUALITY ENGINEERING

The traditional model of segregated teams, which resulted in bottlenecks as well as the glaring absence of interaction, was incompatible with the agile principles. The concept of separated teams was to blame for the bottlenecks in the process. Subsequently, businesses began transitioning to a hybrid model of COE, in which testers engage with developers within sprints while still retaining a shared or independent pool of testers for end-to-end testing, non-functional testing, maintaining common tools, and other tasks with similar responsibilities. Due to its ability to eliminate all the flaws associated with the conventional testing model, accelerate the testing process, and improve the overall quality of the product, the agile software development approach is an excellent fit for the hybrid testing model.

The testing procedures and practices were primarily focused on examining legacy systems. The majority of these systems were desktop-based applications, AS 400s, and mainframes, with certain websites also included. Traditional testing methodologies are no longer sufficient for evaluating modern distributed application architectures due to the wide range of potential combinations of protocols and technologies within a single business transaction. As a result, traditional testing methodologies are inadequate for evaluating modern distributed application architectures. The testing conducted in the past was less technical in nature, with requirements limited to automating regression testing of the user interface, performance testing of significant applications, and test case and defect management. The increasing need for safe and high-performing end-user experiences places significant demands on current business testing methodologies. These methods must be able to keep up with the demand. Both factors are necessary prerequisites for a successful transition to a digital firm. Non-functional testing, which focuses on performance, security, operational acceptability, and user experience, is gaining more attention due to this requirement. Now is the time for quality engineering teams to focus their efforts on API level tests, unit tests, static code analysis, and API level performance testing, among other comparable duties. As a result of this shift in emphasis, a brand-new job title was created: T-Shaped or Software Development Engineer for Testing (SDET). This designation was given to software professionals who specialize in testing software. SDETs possess a unique combination of skills that allows them to not only participate in the testing of existing software, but also contribute to the creation of software from scratch. This distinction sets SDETs apart from other professionals in the field of software development, as they possess a comprehensive skill set. Consequently, there has been an increase in demand for the services provided by SDETs.

The majority of available toolkits in the past were desktop-based, requiring integration through third-party plugins or purpose-built plugins. The number of companies selling testing software was limited, with a few large companies like IBM Rational, Mercury (later acquired by HP), and

Segue dominating the software industry. Over time, the focus shifted from specialized testing tools like UI Automation to fully integrated end-to-end quality engineering platforms like Tricentis TOSCA. An example of this shift can be seen in the transition from UI Automation to Tricentis TOSCA. While UI Automation was initially intended for tasks like automating the user interface, organizations now prioritize not only open-source tools like Selenium, JMeter, and Appium, but also low-cost cloud-based solutions with a monthly subscription model. These cloud-based solutions are available on platforms such as Microsoft Azure and Amazon Web Services, aiming to reduce capital expenditures.

The Waterfall paradigm of low-frequency release has been abandoned in favor of the agile world of weekly and daily releases, which is another revolution that businesses have enthusiastically embraced. This revolution is taking place in the world of software development. This brand-new world is the result of one revolution among many that have just taken place. Test automation has progressed from simply automating user interface level regression tests to test automation within sprints that targets multiple levels, including databases, application programming interfaces (APIs), user interfaces, and so on. This development is a direct result of the goal stated above. In the past, test automation was constrained to the point where it could only automate user interface level regression testing. Because automating the procedures that are related to the testing life cycle enables large time savings, this component of the testing process is given a great amount of significance, which adds considerably to the emphasis placed on this aspect of the testing process. This group of activities includes duties such as static testing, management of test data and environments, maintenance of test environments, and other related activities. In contrast to the traditional practice of running tests in an attended mode toward the end of the software testing life cycle, testers now run automated tests early in the life cycle on every build unattended. This occurs in contrast to the traditional practice of running tests in an attended mode toward the end of the software testing life cycle. This is done in contrast to the traditional practice of running tests in an attended mode at the conclusion of the software testing life cycle. This change was made so that it may be done in place of the traditional practice. On the other hand, the practice of running tests in an attended mode towards the end of the software testing life cycle is something that has been traditionally done. This contrasts with the fact that it is a common practice. With the abundance of mobile and connected devices available on the market, quality engineering teams often find themselves engaged in cross-platform tests that seem to never end. To reduce the amount of time required for testing compatibility across various browsers and devices, engineers are increasingly adopting technologies such as the cloud and containerization.

7.3 ARTIFICIAL INTELLIGENCE (AI) IN QUALITY ENGINEERING

In today's Agile and Digital era, quality engineering organizations must prioritize the redefinition of their enterprise test strategies, implementation of end-to-end test automation, transition towards a hybrid or decentralized model, deployment of SDETs, and enhancement of the skills of their existing staff (Budman et al., 2020). These actions are essential alongside the deployment of SDETs. Only under these circumstances can they thrive and maintain their relevance (Sogeti, 2021).

Artificial intelligence (AI) has transitioned from being a subject of study to impacting our daily lives. This chapter explores how software testing can benefit from the utilization of AI. The advancement of technologies like AI, natural language processing (NLP), and predictive analytics has simplified the creation of more productive and successful applications by software developers (Schatsky & Bumb, 2020). This suggests that AI has the potential to enhance decision-making, increase productivity, improve customer satisfaction, and save costs. When it comes to the topic of AI, testers may question whether AI will replace humans. The answer to this question, both now and in the foreseeable future, is "no." The available AI and testing use cases in the market only complement the expertise of engineers rather than replacing them (Volk, 2021).

You may have contemplated this question during your career as a tester: why don't developers test their own code more frequently? The main reason for this is the lack of time available for test creation. Writing a single unit test case can take anywhere from 8 to 12 minutes of a developer's time (Sugali, 2021). To achieve sufficient test coverage, developers may need to create hundreds or even thousands of test cases. Due to time constraints, engineers are unable to complete their work within the timeframe of agile sprints, which typically last between three and four weeks. As a result, the responsibility for defect identification is largely delegated to the team's test engineers (Sogeti, 2021).

An AI-based tool called DiffBlue can analyze the source code of a project, including its dependencies, and generate a map of the classes and methods involved. Subsequently, a unit test case is created for each method in the application code. These findings are then subjected to a series of checks to ensure their accuracy. The results of the test runs are taken into account when developing the final set of test cases. This greatly aids in expanding the scope of functionality tested by the unit. However, it is important to note that only a small portion of time is dedicated to studying for both exams simultaneously. This serves as an excellent example of how AI can enhance the effectiveness of test design (Alter, 2022).

White box testing is an effective method for identifying bugs in software, ultimately contributing to overall program improvement. It also helps reduce the overall cost of quality (KPMG, 2016).

One of the most crucial opportunities available to a tester is expediting the transmission of feedback to the development team, particularly in terms of testing the build's quality. Let's imagine for a moment that we had a full hour to provide feedback to the app's developer on the version sent for testing. Is such a scenario even feasible? Thanks to the autonomous bot Certifaya, which is capable of recognizing both functional and non-functional behavior, this is now possible. Certifaya checks mobile applications on a wide range of genuine devices, utilizing AI to crawl through apps, identify processes, and verify the operation of each screen. These examinations are not pre-prepared; they are developed in real-time as the process progresses. To complete the exam, the tester only needs to provide credentials, such as a login or password (Budman et al., 2020). Moreover, the bot also provides a wealth of information in the form of videos and images of the evaluated apps. The generated reports take into account various events, including crash data, memory leaks, excessive battery usage, and more. The development of features that enable robots to operate independently is a key focus for many product companies. We anticipate an increased utilization of these capabilities in the coming years (KPMG, 2016).

Do you occasionally review the production tickets submitted by your application support teams? Similarly, do you pay attention to customer feedback about your product on social networking sites like Twitter or Facebook? Testers typically respond with a resounding "no" when asked this question, which is appropriate given the context. After the warranty period expires, there is a tendency to forget about the applications that have been deployed to production. AI can be utilized in various ways to enhance the overall quality of an application. Testing professionals can leverage input to modify the test plan created for your application or product. One way to improve defect analysis is by using classification algorithms that can identify defects based on the functionality of your applications. When a problem is identified in production, algorithms can help determine which developer or support engineer is best suited to address the issue promptly. It is also possible to conduct a high-level root cause analysis of a manufacturing issue and identify the phase at which the fault was introduced into the product (KPMG, 2020). This can be achieved by following the steps outlined in the previous sentence. Feedback from end users can also be segmented based on different aspects of the program using classification techniques and algorithms. Feeds from social media platforms such as Facebook and Twitter can be utilized to gather feedback from customers. If you have a mobile app, the App Store is an excellent platform to receive feedback from app users. Another approach to analyze customer feedback is through the implementation of customer sentiment analysis algorithms. These algorithms estimate the level of satisfaction experienced by your customers. Additionally, closely monitoring the comments and complaints received by your support team is a great way to identify potential issues and prevent them from escalating (KPMG, 2016).

You may wonder why thousands of tests are necessary to validate a small code adjustment. The reason behind this is that the development team may not accurately anticipate the consequences of their code modifications. The use of DevOps has led to an increase in the number of builds expected during a typical development sprint. Consequently, all test cases are executed at the beginning of each build, resulting in a significant amount of time and effort spent on testing. Therefore, it is worth considering an alternative to risk-based testing. If risk-based testing is employed, there is always a possibility that none of the tests will uncover any issues. In such cases, it is possible that problems may slip through into production. On the contrary, Sealights utilizes AI and predictive analytics to analyze the application code and determine the potential impact of changes before suggesting tests (Schatsky & Bumb, 2020). Thanks to the implementation of AI technologies, Sealights is capable of continuously mapping billions of correlations between different tests and approaches. These tests can be either manual or automated, covering unit, integration, API, or regression testing. The platform offers a wide range of alternatives, even allowing for the development of new test concepts. Additionally, based on the provided data, the platform can provide guidance on the type of testing that should be conducted (KPMG, 2020). Furthermore, it can assist in identifying the next steps required to validate the impact achieved by the development team. Consequently, this approach helps reduce the overall time and cost required for the testing cycle by conducting fewer tests. It is important to note that this does not compromise the quality significantly. By utilizing AI and Analytics, tests are conducted in a highly scientific manner.

7.4 CONCLUSION

A significant challenge that arises in the context of digital transformation is the need to reduce the time required to bring high-quality software solutions to market. Businesses strive to meet customer needs while maximizing the accessibility of their products and services. The traditional model of segregated teams, which led to bottlenecks and lack of interaction, is incompatible with agile principles. As a tester, one of the most important opportunities available to you is to expedite the transmission of feedback to the development team. An AI-based tool can analyze the source code of a project, along with its dependencies, and create a map of the classes and methods involved. Non-functional testing focuses on aspects such as performance, security, operational acceptability, and user experience. SDETs possess a unique combination of skills that enable them to participate in both testing existing software and creating software from scratch. It is crucial to stay updated on the latest advancements in the AI industry, as there is significant progress in this field. It is reasonable to expect that in the near future, numerous AI-powered platforms and solutions will emerge to revolutionize software quality engineering.

KEYWORDS

- **algorithms**
- **artificial intelligence**
- **digital transformation**
- **machine learning**
- **mainstream practices**
- **quality engineering**
- **Sealights**
- **test automation**

REFERENCES

Alter, S., (2022). Understanding artificial intelligence in the context of usage: Contributions and smartness of algorithmic capabilities in work systems. *International Journal of Information Management*, *67*, 102392.

Batarseh, F. A., Freeman, L., & Huang, C. H., (2021). A survey on artificial intelligence assurance. *Journal of Big Data*, *8*(1), 1–30.

Black, D. R., Davenport, J., Olszewska, D. J., Rößler, D. J., Smith, A. L., & Wright, J., (2022). *Artificial Intelligence and Software Testing: A Practical Guide to Quality*. BCS, The Chartered Institute for IT.

Budman, M., Bordoloi, S., & Gangopadhyay, N., (2020). *Thriving in the Era of Pervasive AI*. Available online: https://www2.deloitte.com/content/dam/Deloitte/cn/Documents/about-deloitte/deloitte-cn-dtt-thriving-in-the-era-of-persuasive-ai-en-200819.pdf (accessed on 09 March 2024).

Harman, M., (2012). The role of artificial intelligence in software engineering. In: *2012 First International Workshop on Realizing AI Synergies in Software Engineering (RAISE)* (pp. 1–6). IEEE.

KPMG, (2016). *Trends in Software Testing*. Available online: https://assets.kpmg/content/dam/kpmg/pdf/2016/07/Trends-in-software-testing-19july16.pdf (accessed on 09 March 2024).

KPMG, (2020). *Trust in Artificial Intelligence: A Five Country Study*. Available Online: https://home.kpmg/au/en/home/insights/2021/03/artificial-intelligence-five-country-study.html (accessed on 09 March 2024).

Lambert, M., (2021). *How AI is Making Test Automation Smarter*. Available online: https://applitools.com/blog/how-ai-makes-test-automation-smarter/ (accessed on 09 March 2024).

Lenz, A. R., Pozo, A., & Vergilio, S. R., (2013). Linking software testing results with a machine learning approach. *Engineering Applications of Artificial Intelligence*, *26*(5, 6), 1631–1640.

Li, J. J., Ulrich, A., Bai, X., & Bertolino, A., (2020). Advances in test automation for software with special focus on artificial intelligence and machine learning. *Software Quality Journal*, *28*(1), 245–248.

Schatsky, D., & Bumb, S., (2020). *AI is Helping to Make Better Software*. Available Online: https://www2.deloitte.com/us/en/insights/focus/signals-for-strategists/ai-assisted-software-development.html (accessed on 09 March 2024).

Sogeti, (2021). *AI is Becoming Essential for Quality Engineering*. Available online: https://www.sogeti.com/ai-for-qe (accessed on 09 March 2024).

Sugali, K., (2021). *Software Testing: Issues and Challenges of Artificial Intelligence & Machine Learning*.

Volk, T., (2021). *Disrupting the Economics of Testing Through AI*. Available online: https://applitools.com/disrupting-testing-through-ai/ (accessed on 09 March 2024).

CHAPTER 8

India's Digital Financial Inclusion to Digital Adoption: A Memorable Journey

SULAGNA DAS[1] and ABHIJIT DUTTA[2]

[1]*Associate Professor, JIS University, Kolkata, West Bengal, India*

[2]*Professor, Sikkim University, Gangtok, Sikkim, India*

ABSTRACT

With the aim of providing accessible, safe, and transparent financial services to underprivileged and unbanked individuals, Digital Financial Services (DFS) serves as the foundation of Digital Financial Inclusion (DFI) in India. Through this program, consumers will have convenient access to digital banking and financial services via devices, retail agents, and digital transactional platforms. The benefits include quick and easy access to financial resources for all, cost-effectiveness, tailored financial solutions, user-friendliness, and economic empowerment for women and other vulnerable groups in society. The Indian government's demonetization efforts in 2017 led to the creation of a digital economy, resulting in increased usage of digital payment services. However, challenges such as limited internet access, identity theft risks, and a lack of digital transaction point-of-sale devices persist. Utilizing digital banking systems and wallets is a prerequisite for digital financial services, and financial literacy is crucial for effective financial management and informed decision-making. While India's digital adoption score remains lower than that of the UK, South Korea, Sweden, and Singapore, efforts are needed to establish digital financial service infrastructure, particularly in rural and isolated areas with limited financial distribution networks, to ensure widespread mobile connectivity. The Indian government has been promoting online payments since 2016, driven by the 'Digital India' plan. The

Contemporary Digital Transformation and Organizational Effectiveness in Business 4.0.
Nilanjan Ray, Sudin Bag, & Tapas Kumar Chatterjee (Eds.)

Universal Payment Interface (UPI) is the most widely used platform. India is transitioning into a cashless economy by equipping itself with specialized knowledge of digital platforms and financial literacy.

8.1 INTRODUCTION

Digital financial services (DFS) are the core of digital financial inclusion (DFI) in India. DFI relies not only on the digital capabilities of rural customers, but also on how efficiently they conduct their digital financial transactions using their phones and online platforms. The goal of DFI is to utilize digital tools to connect with underserved and unbanked communities, offering them a range of formal financial services and products that are affordable, secure, and transparent, while also meeting their specific needs. Additionally, it promotes effective interaction and networking among individuals.

Any digital financial service is essentially built upon three elements: devices, retail agents, and digital transactional platforms (Lauer & Lyman, 2015). The rapid availability of these three components has created a favorable environment for the exponential growth of access to digital banking and financial services. According to BRICS Digital Financial Inclusion Report (2021), the benefits of DFI include the quick and widespread delivery of financial services, universal access to financial resources, cost-effectiveness of digital platforms compared to physical ones, accessibility to personalized, need-based financial products and services at reduced costs for a wide range of customers, ease of use, reduced risks and costs associated with handling physical cash, and economic empowerment of women, youth, vulnerable groups, and marginalized sections of society.

The benefits of DFI also come with potential threats and challenges. Technology-driven financial activities may be significantly excluded (Dewan & Chen, 2005), worsening the current digital divide due to issues of consumer data security and an increase in cybersecurity incidents. If these issues are not resolved, the advantages of digitalization may be limited. However, the overall benefits of initiatives towards DFI outweigh any potential drawbacks. These initiatives are guided by smart regulations that address the associated risks, enabling many individuals and businesses to participate in the formal financial system.

8.2 DEMONETIZATION IN INDIA

In November 2016, Prime Minister Mr. Narendra Modi announced 'demonetization' as a measure to eradicate black money from the country (Srivastava &

Raghubir, 2008). This action led to the discontinuation of currency notes with denominations of Rs. 500 and Rs. 1,000. As a result, 86% of India's currency was rendered worthless overnight. The value of currency notes declined significantly from 17.97 trillion to 7.8 trillion by January 2017, and 99% of the demonetized currencies were returned to the Reserve Bank of India (business-standard.com). However, despite these outcomes, demonetization had a positive impact on India's economy (Rathore, 2016). It resulted in increased tax collection, greater reliance on digital currencies and payment platforms such as UPI and digital wallets, and efforts to combat black money and reduce human trafficking.

8.3 IMPACT OF DEMONETIZATION ON DIGITALIZATION

Following the demonetization initiatives of the Government of India (GOI) in 2017, the usage of digital payment platforms such as digital banking and payment gateways has gained momentum. The Prime Minister, in several of his speeches and interviews, has shown a keen interest in making India a digital economy (Dezan Shira and Associates, 2017). The forefront of this economy would be the usage of digital banking and payment gateways.

A digital economy enabled by digital banking and payment gateways has several benefits. India's cash to GDP ratio has ranged from 8.8% in March 2017 to 11.4% in March 2019. The currency in circulation has increased by 1.5% for every percent increase in GDP since 2019. This is happening despite the GOI's efforts to promote a less cash economy. A HSBC report on the Indian economy (HSBC Report on Indian Economy, 2018) has pointed out that the use of higher cash could be due to the easing out of GST and the benefits flowing downstream.

The GOI has implemented several initiatives to promote a cashless economy. Prominent among them are – (i) united payment interface (UPI); (ii) unstructured supplementary service data (USSD) developed by NPCI using GSM community technology without an internet interface; (iii) Bharat interface for money (BHIM), a UPI platform launched by the GOI on December 30th, 2016; (iv) Aadhaar-enabled payment system (AEPS); and (v) e-payments using debit and credit cards.

Digital banking and payment gateways have several advantages, including: (a) transparency in the amount held in banks by individuals, resulting in increased tax collection; (b) reduction of red tape and bureaucracy; (c) significant reduction in operational costs for banks; (d) extensive bank reach, leading to greater inclusion; and (e) ease in service delivery by financial institutions (Pratik, 2017). However, there are hurdles and disadvantages of digital financial platforms in India. The foremost among them are the non-availability of consistent internet facilities, fear and risk of theft of personal information on digital platforms, and the non-availability of Point of Sale (POS) machines for digital service payments.

Despite these challenges, digital payments have gained momentum in India. According to data from NITI Ayog, the total digital transactions have increased from 24.2% in the year 2016–2017 to 55% in the recent years (Watal, 2017). However, there has been a recent decline. RBI data shows that the total volume of digital transactions reached Rs. 149.58 billion in 2017 and increased to 115490.30 billion in February 2019. Figures 8.1–8.3 give us a clear image of the impact of demonetization on digitalization.

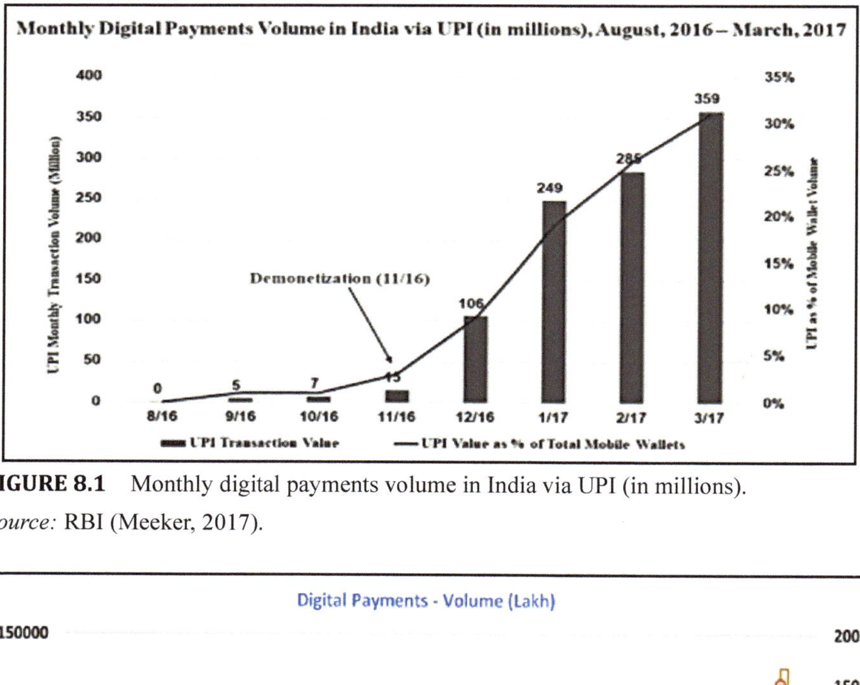

FIGURE 8.1 Monthly digital payments volume in India via UPI (in millions).

Source: RBI (Meeker, 2017).

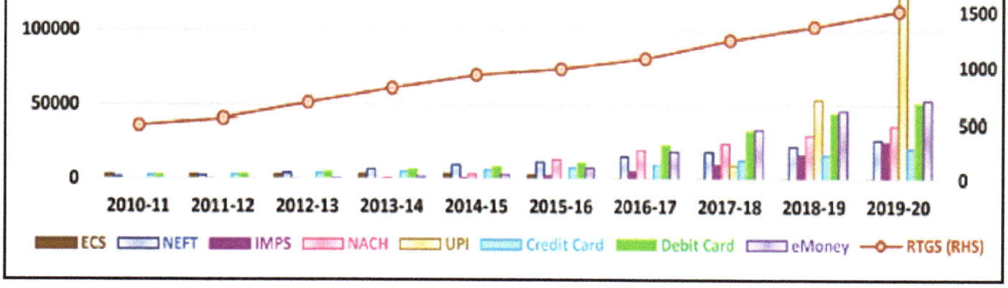

FIGURE 8.2 Digital payment system in India volume.

Source: RBI data.

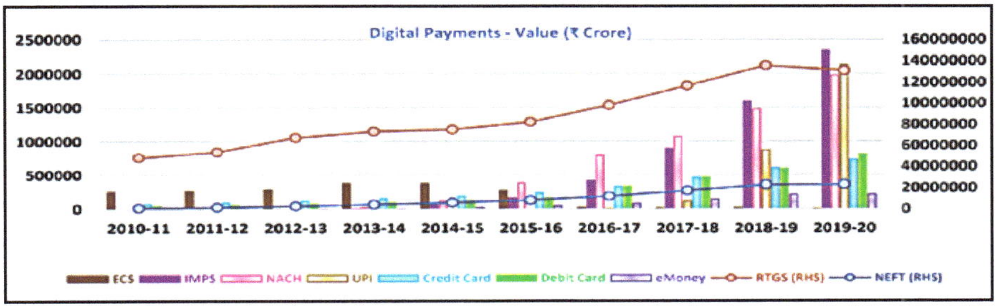

FIGURE 8.3 Digital payment system in India value.

Source: RBI data.

Figures 8.1–8.3 give an idea of the monthly volume of digital payments, both before and after demonetization, in terms of value and volume. It is evident that the use of digital payments has increased in the post-demonetization period, as indicated by the upward trend in the curve. Figure 8.1 shows a relatively higher increase in the use of mobile payments during the post-demonetization era.

8.4 OBSTACLES IN THE PATH OF DIGITAL FINANCIAL INCLUSION (DFI)

The road to digitalization is not simple in a country like India, where many individuals lack bank accounts and widespread Internet access. There are numerous obstacles to overcome in order to achieve the goal of a cashless economy. To reach the masses, innovative thinking and effective techniques must be devised. It is not feasible for the government to instill digital habits among those who still reside in rural areas and lack basic amenities (Bhatnagar, 2017). Despite the Indian Government's efforts to establish interconnected digital infrastructure, the acceptance of DFS in rural areas is hindered by digital financial illiteracy. The lack of trust in digital technology, absence of rural customer-centric financial products and services, inability to use smartphones, and poor network connectivity prevent the rural population from embracing digital methods of payment (Modi, 2021).

Consumers who possess digital financial literacy and awareness are better equipped to make informed decisions regarding borrowing, investing, and day-to-day money management (Capuano & Ramsay, 2011). Consequently, there is a direct correlation between digital financial literacy and digital inclusivity. This is particularly crucial in developing countries, where efforts to combat poverty and improve welfare are impeded by a lack of financial literacy. Depending on one's understanding of financial products, their financial situation can vary significantly (Subha & Priya, 2014).

In many developing countries, agriculture employs a significant portion of the population. These regions are particularly susceptible to revenue shocks caused by weather and inflation-related issues, as well as resulting price volatility in the goods they produce. To cope with such shocks, households may need to heavily rely on their savings and insurance. However, low levels of digital financial literacy pose an additional barrier to digital financial inclusion (DFI). These obstacles are influenced by various factors, including the environment, financial provider profiles, and customer segments. They span from the demand side to the supply side. For underserved populations, barriers include issues such as financial scarcity, behavioral biases, and limited financial literacy (Refera, Dhaliwal, & Kaur, 2016). On the market side, common barriers include policy and regulations, physical and infrastructure constraints, unsuitable products, and inappropriate sales practices (Grohmann & Menkhoff, 2020).

A potential customer's adoption or use of a financial service or product may be negatively impacted by a lack of understanding about the service or product. The inability to compare the tariffs of various financial service providers, high transaction costs, incomplete information about tariffs, and a lack of regulations governing pricing and tariff transparency may all deter customers from using financial services. In addition to having a higher likelihood of being financially excluded than people with the highest levels of education, low-income segments and disadvantaged groups like women, young people, and people who have been forcibly relocated often fall into these categories (torontocentre.org). With respect to DFI specifically, factors like a consumer's inability to navigate a mobile money menu or read financial text messages (including mobile money confirmation or promotional messages and bank balances) due to a lack of financial proficiency, it has become challenging for them to access mobile banking and financial services or make financial decisions. Additionally, it puts consumers at risk for fraud, such as when they are unable to view the balance in their mobile wallet (Lusardi, 2012).

Digital financial literacy and the adoption of digital financial products and services may also be hindered by a number of other factors, including poverty, misunderstandings, and mistrust of financial services, a lack of formal education, providers' inappropriate marketing strategies that exclude certain demographic groups, a lack of price transparency, and unfair market practices. People in rural areas or with low incomes who are not financially literate may also feel pressured or discouraged from utilizing DFS like mobile money. They might think that products like mobile financial services are too complicated for them to understand, which would prevent them from using such products (Toronto Center, 2019).

8.5 OVERCOMING THE OBSTACLES OF DIGITAL FINANCIAL INCLUSION (DFI) WITH THE HELP OF DIGITAL FINANCIAL LITERACY

According to the definition of financial literacy, it is "a mix of financial awareness, knowledge, understanding, skills, attitude, and behaviors necessary to make wise financial decisions and manage one's own finances, and ultimately attain individual financial well-being." Digital Financial Literacy is the combination and application of both financial and digital literacy to facilitate the use of DFS, through digital banking platforms as well as digital wallets, and also handle the financial risk involved in these transactions. The definition is well explained by the diagram in Figure 8.4.

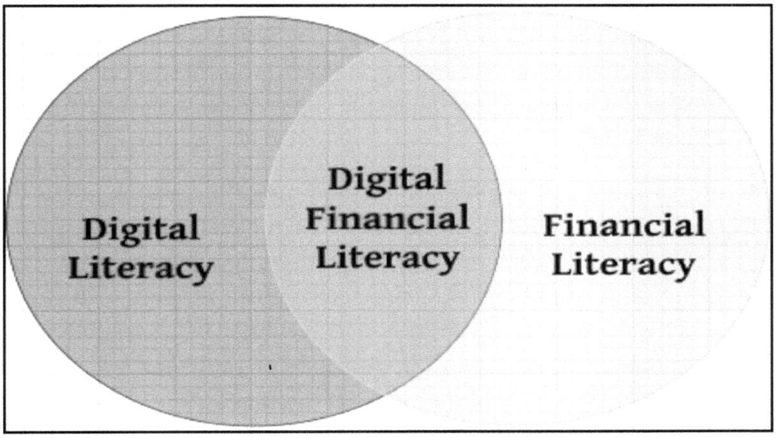

FIGURE 8.4 Digital financial literacy.

Source: Drawn by Author.

The need of the digital era is for everyone to be digitally and financially literate, enabling them to effectively plan their finances, retirement savings, and pension plans (OECD, 2019). Consumers should possess a comprehensive understanding of financial knowledge and concepts in order to utilize financial technology (fintech) products and services efficiently, while also safeguarding themselves against financial fraud and costly mistakes.

Digital financial literacy encompasses four dimensions:

1. **Knowledge Level:** The first dimension is knowledge level, which involves understanding digital financial products and services (DFS), such as digital payments (electronic fund transfers, e-wallet payments, and remittance services) and asset/insurance management (online trading, managing demat accounts, and online insurance premium payments).
2. **Digital Financial Risks:** The second dimension, digital financial risks, pertains to the attitude towards the additional financial risks associated with digital platforms, including awareness of phishing, pharming, spyware, hacking, and other threats (Morgan, Huang, & Trinh, 2019).
3. **Digital Financial Risk Controls:** The third dimension, digital financial risk controls, focuses on the skills required to identify risks, protect against fraud and losses, and navigate redress procedures.
4. **Consumer Rights and Redress Procedures:** The fourth dimension, consumer rights and redress procedures, addresses the knowledge of how and where to file a claim for compensation in the event of a loss due to fraud.

Therefore, having a strong understanding of digital financial literacy can help consumers become more aware of the various kinds of financial products and services that are available digitally, and to have sufficient knowledge to understand which characteristics are most crucial to take into account when selecting such products and services. Digital financial literacy can give consumers the confidence to make an informed choice about using a traditional or new type of financial service, such as mobile money and digital savings and credit, once they have a better understanding of the financial products available to them and the necessary skill set to use those products. Once they have the requisite financial skills, they feel confident about making financial decisions, and they no longer have any negative thoughts of transacting in digital mode. Digital financial literacy contributes to the efficiency, transparency, faith, trust, and protection of the consumers who access the digital platforms for transactions. Higher levels of literacy could also enhance consumer protection, the safety and soundness of financial transactions through financial institutions, confidence among the users, and fair, efficient, and transparent markets, ensuring secure access and market development.

The success of DFS depends on the shift from cash to a digital payments' economy in rural India, which can only be achieved when the rural population is digitally and financially literate. In the current COVID-19 situation, Digital Financial Literacy (DFL) has become an increasingly important aspect of education and a powerful tool for DFI (Modes of Digital Payment, n.d.). DFL refers to the knowledge of digital financial products and services, awareness of the financial risks involved (World Bank, 2018), the ability to protect oneself from fraud and other losses, and access to procedures for redress in case of falling victim to such risks.

8.6 ATTAINMENT OF DIGITAL FINANCIAL INCLUSION (DFI) THROUGH DIGITAL FINANCIAL LITERACY, DIGITAL AWARENESS, AND DIGITAL ADOPTION IN INDIA

The Ministry of Electronics and Information Technology, GOI, has implemented numerous initiatives and promotional schemes to enhance Digital Financial Literacy. Two such programs are Digital Jagriti and DigiShala, which aim to raise awareness about digital finance among the citizens. The Digital Jagriti program encourages merchants to adopt digital payment methods and facilitates individual usage of digital payments. On the other hand, DigiShala is a free educational channel available in Hindi, English, and regional languages on Doordarshan DTH, providing information about various electronic payment methods. Additionally, there are other schemes such as the BHIM-Cashback Scheme for Merchants, BHIM-Referral Bonus Scheme for Individuals, and the Promotional Scheme of BHIM Aadhaar. The Ministry of Railways offers incentives (Incentives and Promotional Schemes, n.d.) like exemption from service charges on E-Ticket bookings, while the Ministry of Road Transport and Highways provides a 10% discount to toll road users who pay digitally. Awareness programs (Awareness Programs, n.d.) include workshops, video conferencing sessions, and knowledge sessions on digital payments and bill payments for different segments of the population.

Figure 8.5 illustrates the parameters used to measure Financial Inclusion. The three parameters shown are access (including internet connection and speed), quality (covering digital financial literacy and grievance handling), and usage (referring to the adoption of DFS through digital platforms).

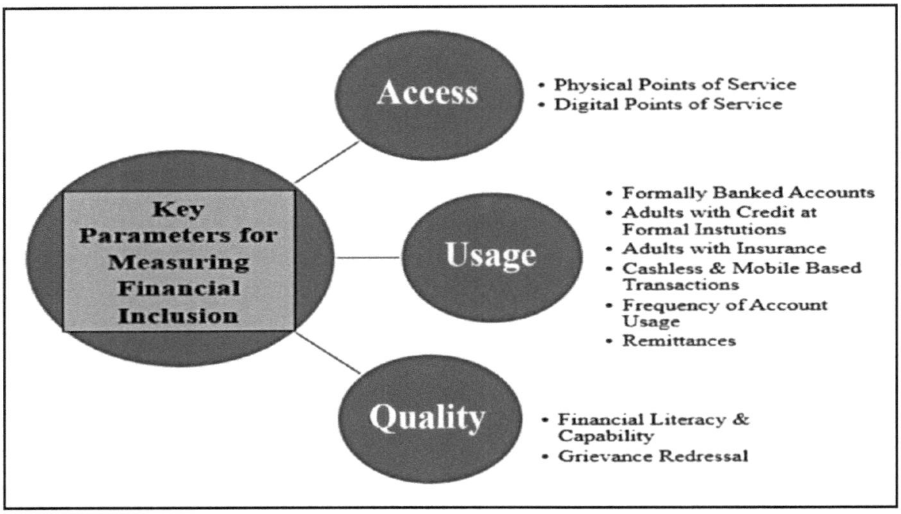

FIGURE 8.5 Measuring financial inclusion.

Source: RBI (NSFI).

A country's digital adoption index is based on three factors: digital foundation (i.e., cost, internet speed, and reliability of the internet connection), digital reach (number of mobile devices, applications downloaded, and usage of data), and digital value (customer engagement in chatting, making digital payments using digital platforms) (Digital India, 2019). Digital adoption index of India is further shown by Figures 8.6 and 8.7.

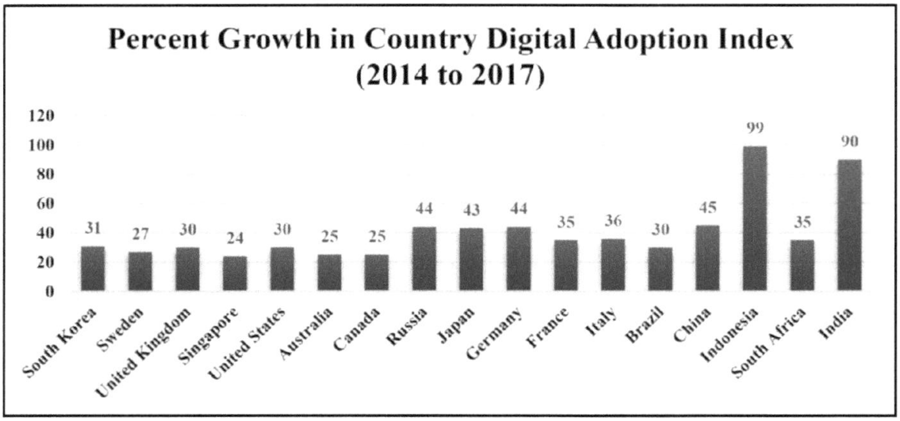

FIGURE 8.6 Percent growth in country digital adoption index (2014 to 2017).

Source: McKinsey Global Institute (2019).

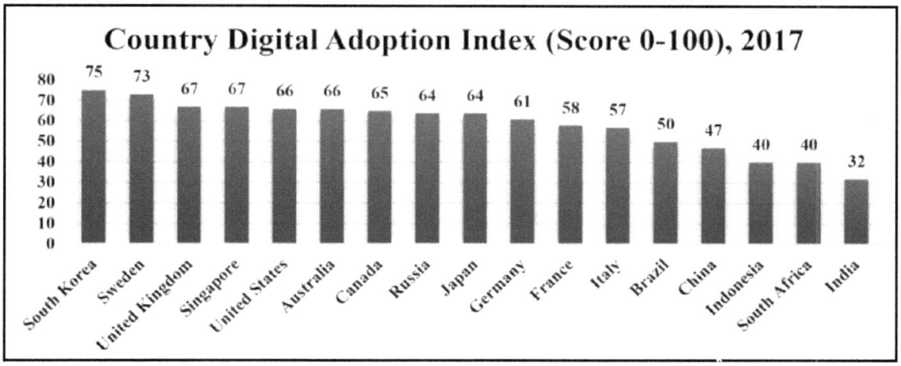

FIGURE 8.7 Country digital adoption index (score 0–100), 2017.
Source: McKinsey Global Institute (2019).

According to a report (mckinsey.com) by the McKinsey Global Institute, India's score rose by 90% between 2014 and 2017, second only to Indonesia's improvement, at 99%, over the same period (Figure 8.6). In absolute terms, India's score is low, at 32 out of a maximum of 100, comparable to Indonesia's at 40 (Figure 8.7), but significantly lags behind the four most digitalized countries, namely South Korea, Sweden, Singapore, and the United Kingdom.

8.7 DEVELOPING DIGITAL INFRASTRUCTURE FOR SUPPORTING DIGITAL FINANCIAL INCLUSION (DFI) IN INDIA

Building the necessary infrastructure for the expansion of DFS is crucial for any nation. The adoption of DFS relies on having the digital infrastructure required for widespread mobile access and usage, especially in rural and remote regions of India where financial distribution networks are limited and mobile network coverage is inadequate. To expand DFS and reach a larger portion of the population in rural, semi-rural, and urban areas, it is equally important to have a financial infrastructure that supports effective, affordable, and secure transactions. This includes interconnected digital payment systems and distribution networks, digital identification, and reliable security platforms (adfi.org).

The essential components of DFI infrastructure are digital transactional platforms (such as Digital Wallet and UPI) that enable clients to use devices for transmitting and receiving transaction data and connecting to authorized banks or non-banks to make or receive payments, transfers, and store value electronically. Customers can interact with these devices using smartphones or analog instruments like debit/credit cards, including point-of-sale (POS) terminals. Additionally, customers can convert cash into electronically stored value and vice versa by utilizing retail agents equipped with digital devices connected to communication infrastructure for transmitting and receiving transaction details. Furthermore, banks and non-banking financial institutions can offer additional financial services like credit, savings, insurance, and securities through digital platforms, providing extra benefits to underprivileged and unbanked individuals. This encourages reliance on digital platforms and motivates them to engage in mobile banking transactions (Digital Financial Inclusion, 2015).

Building a reliable digital financial infrastructure with a strong internet connection and high internet speed across India is a challenging task. However, according to the press release from TRAI (The Indian Telecom Services Performance Indicator Report, 2022), India is making great progress in achieving its 'digital reach' target. The country is on the verge of a new revolution, thanks to its increasing reliance on digital financial technologies and digital payment platforms.

Currently, India has a total of 1,167.82 million telephone subscribers, with 7.82 million subscribers requesting Mobile Number Portability. The urban subscribers amount to 646.99 million, while the rural subscribers reach 520.82 million. Additionally, there are 788.77 million broadband subscribers. Figures 8.8 and 8.9 illustrate the mobile subscription from 2000 to 2021 and the internet subscription from 2014 to 2021.

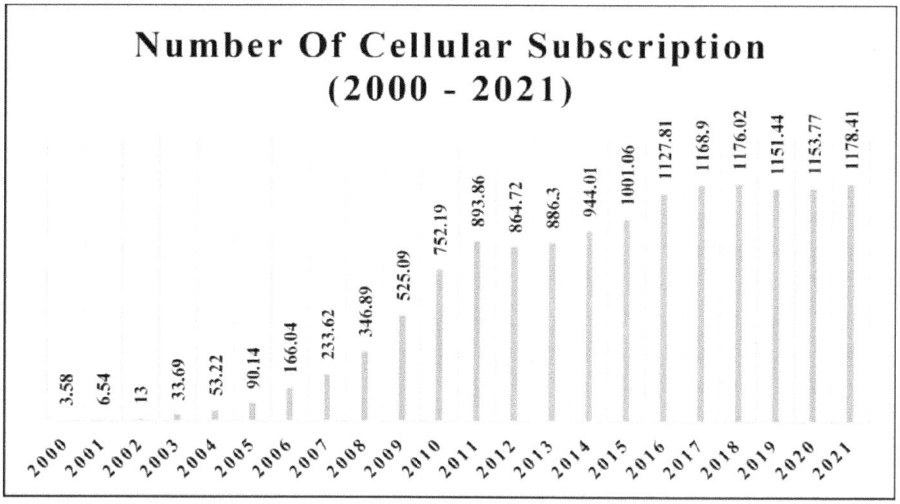

FIGURE 8.8 Mobile subscription in India (in millions) (2000–2021).

Source: Telecom Statistics India (2020); compiled by the author.

Figure 8.8 shows an increasing trend in mobile subscriptions in India. It began with only 3.58 million in the year 2000 but has since grown to 1178.41 million in 2021. It is evident that India has successfully developed the digital infrastructure necessary for DFI.

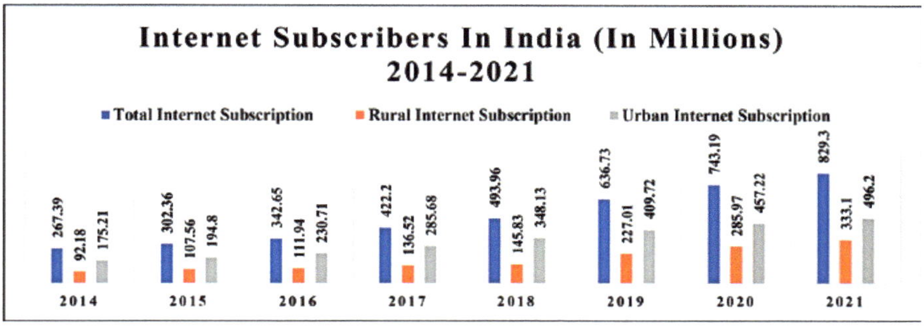

FIGURE 8.9 Internet subscribers in India (2014–2021).

Source: TRAI reports (2022); compiled by the author.

Figure 8.9 illustrates a rising trend in internet subscriptions in India. In 2014, there were only 267.39 million subscribers, but by 2021, this number had increased to 829.3 million. The data also indicates a rising trend in both urban and rural subscribers. This suggests that the people of India are actively embracing digitalization and are familiar with various digital applications and platforms in their daily lives.

8.7.1 DIGITAL PAYMENT INDEX BY RESERVE BANK OF INDIA

The Reserve Bank of India has created a composite digital payments index (DPI) to assess the status of the country's digital payments and transactions. This index was established in 2018.

To measure the extent and penetration of digital payments in the country across various time periods, five parameters were developed, each with different weights as explained in Figure 8.10. Each of these five parameters consists of five to six sub-parameters, which serve as individual measurement indicators. The parameters are discussed with Figure 8.10.

The following criteria apply: (i) payment enablers (weight 25%), with sub-parameters internet, mobile, Aadhaar, bank accounts, participants; and merchants; (ii) payment infrastructure – demand-side factors (10%), with sub-parameters debit cards, credit cards, prepaid payment instruments,

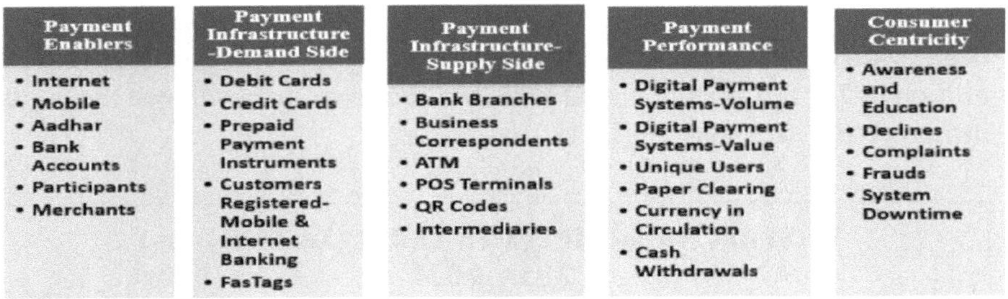

FIGURE 8.10 Reserve Bank of India – digital payment index – parameters and sub-parameters.

Source: RBI website.

customers registered for mobile and internet banking, and Fastags; (iii) payment infrastructure – supply-side factors (15%), with sub-parameters bank branches, business correspondents, ATMs, POS terminals, QR codes, and intermediaries; (iv) payment performance (45%), with sub-parameters digital payment system volume, digital payment system value, unique users, paper clearing, currency in circulation, and cash withdrawals; and (v) consumer centricity (5%), with parameters awareness and education, declines, complaints, frauds, and system downtime. The sub-parameters themselves consist of different quantifiable indications. Each parameter has its own key sub-parameters provided. The digital payment index for September 2021 (Reserve Bank of India Announces Digital Payments Index) stands at 304.06, compared to 270.59 for March 2021.

8.8 IMPACT OF COVID-19 ON DIGITAL PAYMENTS

The government has been actively promoting online payments since 2016, starting with demonetization. The initiative of 'Digital India' has been the driving force behind these decisions. India offers various methods of digital payments, and to encourage people to embrace digitalization, the government has provided incentives for digital payments (Adhikari et al., 2022). Prior to the pandemic, mobile payments were not widely used. However, during the pandemic and lockdown situation, there has been a significant growth in the use of digital payments (Achutamba & Hymavathi, 2022). As the number of COVID positive cases increased, people became hesitant to use cash and started viewing money as 'unclean' and 'risky,' leading them to prefer cashless transactions. The fear of the coronavirus spreading through touch prompted more people to choose digital payments over cash or traditional options (Singhal & Gupta, 2021). The COVID situation necessitated a shift towards DFS to ensure uninterrupted financial transactions. The IMF Financial Access Survey (2021) collected data on two types of DFS: mobile money (mobile wallets, UPI, Paytm) and mobile and internet banking. Among these, mobile money was more prevalent in developing economies like India, while mobile and internet banking was preferred in developed economies. The pandemic has compelled both individuals and businesses in India to reconsider their payment platforms and methods, incorporating factors that facilitate hassle-free financial transactions.

Today, a majority of the population in India has opted for online or digital payments, utilizing platforms such as UPI and mobile wallets. The National Payments Corporation of India (NPCI) (NSFI, 2020) reported that in September 2020, there were over 1.8 billion transactions on the unified payments interface (UPI), compared to 1.61 billion transactions in August 2019. Transactions conducted through the immediate payment service (IMPS) amounted to 27.9 million, while 23.1 million transactions were made using Bharat Bill Pay. The Impact of the COVID-19 Outbreak on Digital Payments (n.d.). According to data from bankers, there was a significant decline in high-value spending on hospitality and travel in 2020 due to travel restrictions. However, by September 2020, the volume of transactions on e-commerce websites using credit and debit cards had returned to pre-COVID levels.

A study conducted by Statista on the impact of the coronavirus (COVID-19) on digital payment app usage in India in 2020 (COVID-19 Impact on Digital Payment App Usage in India) revealed the percentage of digital platforms used by people during the pandemic (*see* Figure 8.11).

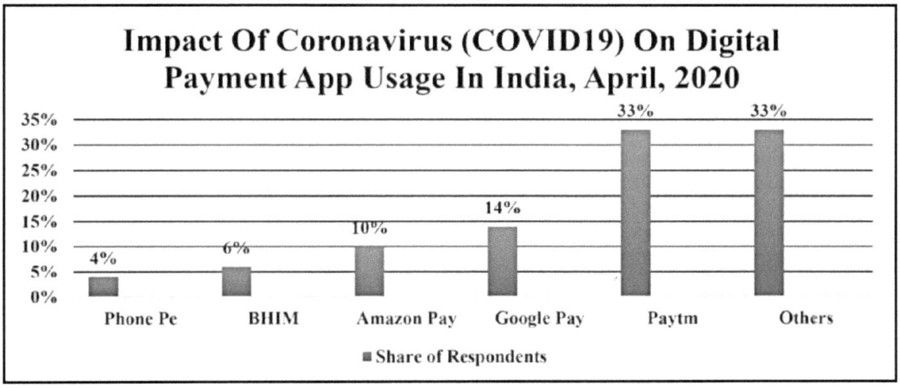

FIGURE 8.11 Impact of COVID-19 on digital payment in India.

Source: Statista (compiled by the author).

The results of this survey conducted in India on the impact of the coronavirus (COVID-19) and subsequent lockdown reveal that a majority of respondents reported using Paytm for digital payments. More than 30% of the survey participants noted an increase in the use of online payments. While COVID-19 has played a significant role in reducing cash transactions, it has not completely eliminated them. However, the pandemic has contributed to the growing acceptance of digital payments in India.

One of the major barriers to the adoption of digital payments in India was the complexity of the existing payment infrastructure. However, the disruption caused by the pandemic provided businesses with an opportunity to reassess their payment processes and simplify them through digital payment platforms. People have chosen digital currency to avoid physical handling of cash. The survey identified lack of knowledge, awareness, technicalities, unavailability of proper digital infrastructure, and associated costs as the main reasons for the fear of using digital payment platforms (Impact of COVID-19 on Digital Payments in India, 2020).

8.9 AWARENESS AND USAGE PATTERN OF DIGITAL PAYMENT PLATFORMS IN INDIA

The impact of COVID-19 on people's lives has been significant in the last few years. The lockdown period in India has brought about changes in people's payment behavior due to changes in their lifestyle. While most household members in the top and middle segments were at home, those in the bottom section struggled to meet their needs. People have been learning about the uses of digital commerce while also discovering new ways to live, work, and transact.

India has witnessed a significant increase in the usage of digital payment applications for financial transactions, particularly since the pandemic forced people to stay indoors and place orders for food, groceries, and other goods through online marketplaces. Many online retailers and e-commerce platforms have disabled cash payments to minimize interaction with delivery personnel.

According to a report from the Ministry of Electronics and IT, the volume of digital financial transactions has increased by 33% from 2020–2021 to 2021–2022. A total of 7,422 million digital payment transactions were recorded in 2021–2022, compared to 5,554 million digital payment transactions in the year 2020–2021. The most popular digital platform used by people was the Universal Payment Interface (UPI), as reported by the NPCI, and the total number of transactions was nearly double the amount from a year ago. India made 7,422 Crore digital payments in FY22 at a growth rate of 33% (MeitY, 2022). According to records, UPI recorded over 45 million transactions, 2 billion transactions were made through AePS, and 4.6 billion transactions were made through IMPS in the financial year 2022. The usage of digital financial platforms (Volume & Value) in the year 2021–2022 is discussed in Figures 8.12 and 8.13.

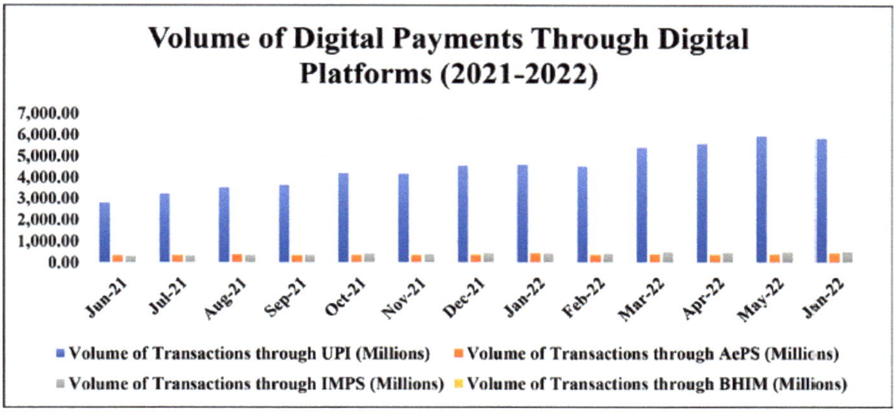

FIGURE 8.12 Volume of digital payments through digital platforms (2021–2022).

Source: NPCI (compiled by the author).

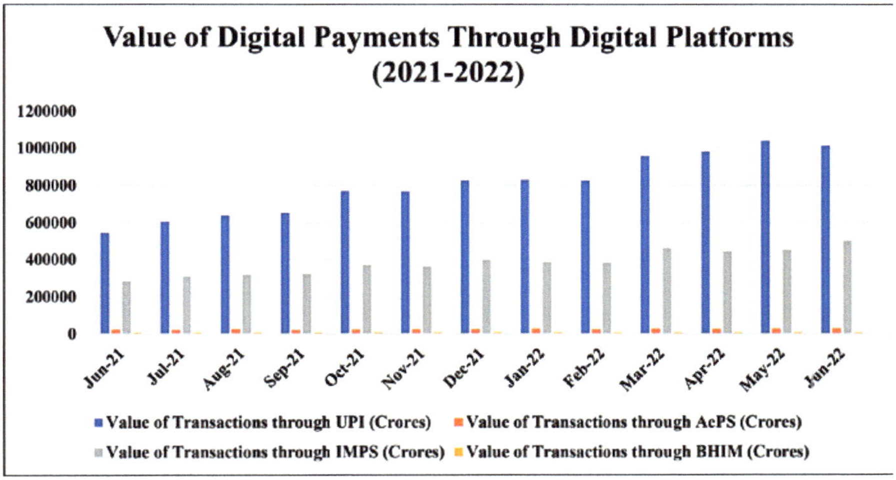

FIGURE 8.13 Value of digital payments through digital platforms (2021–2022).

Source: NPCI (compiled by the author).

From Figures 8.12 and 8.13, it has been observed that the volume and value of UPI transactions were maximum in comparison to the other modes of payments (Modes of Digital Payment, n.d.).

In a survey conducted by NPCI on households (bottom-income group, middle-income group, and top-income group) in 25 states of India, to assess the awareness, adoption, and usage behavior of digital payments, it was observed that almost one-third of the households were using digital payments in some way or another. It is noteworthy that the issue of digital payment is not limited to the wealthy or educated; the bottom 40% also expressed a desire to use digital payments. The availability of smartphones for digital transactions is no longer a barrier for Indians, as the numbers indicate. Digital payments have been widely accepted. Table 8.1 provides a clear overview of the adoption of digital payments by Indian households.

TABLE 8.1 Status of Digital Payment Adoption of the Indian Households

Percentage of Household in Each Income Group Who Have...	Income Groups			Total
	Bottom 40%	Middle 40%	Top 20%	
Never used it	72%	52%	39%	59%
Used in the past but stopped	4%	13%	12%	9%
Use now	24%	34%	49%	32%
Never used but would like to use	16%	14%	9%	14%
Potential user base if				
"Desire to use" are converted	40%	48%	58%	46%
"And past users brought back"	44%	61%	70%	54%
"Have smart phones"	57%	72%	90%	68%

Source: NPCI.

Table 8.2 gives an idea of the digital payment methods used by households. From the data, it is clear that households prefer Paytm, PhonePe, and UPI for digital transactions.

TABLE 8.2 Digital Payment Methods Used by the Households

Percentage of Household Who do Digital Payments in Each Income-Group Using...	Income Groups			Total
	Bottom 40%	Middle 40%	Top 20%	
Paytm, PhonePe type applications	80%	79%	78%	79%
UPI	56%	45%	56%	52%
Credit or debit card	27%	42%	44%	38%
Bank applications	18%	41%	43%	34%

Source: NPCI.

Table 8.3 discusses the online banking and mobile banking applications used by households for mobile banking.

TABLE 8.3 Status of Online Banking – Done by the Households

Percentage of Household in Each Income-Group Who Have...	Income Groups			Total
	Bottom 40%	Middle 40%	Top 20%	
Have smart phone	57%	72%	90%	68%
Do digital payments	24%	34%	49%	32%
Have mobile banking applications	14%	37%	59%	31%
Use phone for banking related matters	5%	17%	30%	14%

Source: NPCI.

Compared to other payment methods, online banking is less developed but has significant potential. In comparison, fewer people are using mobile phones for banking compared to those using mobile banking apps. There is a 36% gap between smartphone owners and users of digital payments. Immediate action should be taken to bridge this gap through financial literacy programs that promote app download and usage (Digital Payments Adoption in India, 2020).

A well-developed ecosystem has created strong consumer momentum in favor of online banking and digital payments. People are well aware of digital payment applications and are actively using these platforms for their financial transactions. Consumer perception of the benefits and drawbacks of digital payment applications indicates a high level of consumer awareness towards digital platforms. India is rapidly moving towards becoming a cashless economy, empowering its citizens with financial literacy and specific knowledge on how to use digital platforms, solve related problems, and safeguard themselves from risks. The majority of people, including those with lower incomes, have now adopted digital platforms.

In the current year 2022, there were 9.36 billion transactions amounting to INR 10.25 trillion processed through various payment modes such as mobile wallets, UPI, prepaid cards, and debit and credit cards. UPI transactions emerged as the most preferred payment mode among consumers, with a market share of 64% in volume and 50% in terms of value (Worldline 'India Digital Payments Report' Q1, 2022). According to NPCI data, PhonePe was the leading UPI app in February 2022, with 212 crore transactions, while Google Pay closely followed with 152.4 crore transactions.

8.10 FUTURE OF DIGITAL PAYMENTS

India's digital payment scenario has undergone a significant transformation over the last five years and is currently in the process of developing a digital payment economy. The country has witnessed a substantial increase in the usage of digital payment applications for financial transactions, especially since the pandemic compelled people to stay indoors and place orders for food, groceries, and other goods through online marketplaces. A report published by PhonePe reveals that people are increasingly embracing digital payment platforms for transactions, and it is projected that within the next five years, two out of three transactions will be conducted digitally, rendering cash less dominant. According to the analysis, non-cash transactions are expected to account for nearly 65% of the total payment volume in the next five years, surpassing 10 trillion dollars, more than triple the current amount (Goyal et al., 2022).

Furthermore, users of the UPI ID can now receive payments from abroad directly into their bank accounts. NIPL (the international arm of NPCI) has entered into an agreement (MoU) with TerraPay to facilitate foreign remittances (Now Indians Will be Able to Receive Money from Overseas via UPI, 2022). As per NPCI, the collaboration with TerraPay (n.d.) will enable UPI users to leverage TerraPay's interoperable payments infrastructure for receiving payments through a fully regulated, secure, scalable, and efficient channel.

This is NPCI's effort to provide a global payment solution that can enhance the usage of UPI. This NPCI initiative will also benefit Indian travelers visiting abroad, as they can utilize UPI for purchases similar to debit or credit cards (Soon, You Can Pay via UPI All Over the World, 2019).

The Ministry of Electronics & Information Technology (MeitY) has organized the Digital Payments Utsav to promote digital payments in the country from December 5th, 2021, to March 5th, 2022. It is a government initiative to expedite the adoption of digital payments, empowering citizens, and merchants with convenient digital payment solutions (Digital Payments Utsav, n.d.). The future of digital payments in India is indeed promising, as the country has introduced various innovative techniques such as "Payments on the Go" with Bank of Baroda, RuPay-on-the-Go launched by City Union Bank, "Inclusive Credit for All" with India Post Payments, and other banks launching contactless credit cards on the RuPay network, etc. (MeitY, 2021). In the near future, the Indian economy will transition towards a cashless society. Countries that once believed they were ahead of India in terms of creativity and technology are now envious of the country's remarkable performance in the digital payment arena.

KEYWORDS

- **demonetization**
- **digital adoption**
- **digital financial inclusion**
- **digital financial literacy**
- **digital payment index**
- **digitalization**
- **Reserve Bank of India**
- **TerraPay**

REFERENCES

Achutamba, V., & Hymavathi, C. H., (2022). Impact of COVID-19 on digital payments in India. *Journal of Positive School Psychology*, *6*(3), 4394–4400.

Adhikari, S., Pallavi, D. R., Ghimire, D., Thapa, S., & Sadikshya, (2022). Impact of COVID-19 on digital payment system of India. *AIP Conference Proceedings*, *2393*(1), 020178. https://doi.org/10.1063/5.0074251.

Awareness Programs, (n.d.). *Awareness Programs.* Ministry of Electronics and Information Technology, Government of India. meity.gov.in (accessed on 09 March 2024).

Bhatnagar, H., (2017). Demonetization to digitalization: A step toward progress. *Management and Economics Research Journal, 3*, 11–15.

BRICS Digital Financial Inclusion Report, (2021). Retrieved from: https://brics2021.gov.in/brics/public/uploads/docpdf/getdocu-53.pdf (accessed on 09 March 2024).

Capuano, A., & Ramsay, I., (2011). *What Causes Suboptimal Financial Behavior? An Exploration of Financial Literacy, Social Influences and Behavioral Economics.* U of Melbourne Legal Studies Research Paper No. 540, Available at SSRN: https://ssrn.com/abstract=1793502 or http://dx.doi.org/10.2139/ssrn.1793502 (accessed on 09 March 2024).

COVID-19 Impact on Digital Payment App Usage in India 2020, (2022). India – COVID-19 impact on digital payment apps 2020. Statista.

Demonetization, 5 Years on: Key Economic Indicators Paint a Mixed Picture, (2021). Business Standard News (business-standard.com).

Dewan, S. G., & Chen, L. D., (2005). Mobile payment adoption in the USA: A cross-industry, cross-platform solution. *Journal of Information Privacy and Security, 1*, 4–28.

Dezan Shira and Associates, (2017). *Growth of Digital Payments Systems in India.* http://www.india-briefing.com/news/growth-of-digital-payments-systems-in-india-14797.html/ (accessed on 09 March 2024).

Digital Financial Inclusion, (2015). worldbank.org.

Digital India Technology to Transform a Connected Nation, (2019). Retrieved from: https://www.mckinsey.com/capabilities/mckinsey-digital/our-insights/digital-india-technology-to-transform-a-connected-nation (mckinsey.com).

Digital India, (2019). https://datareportal.com/reports/digital-2019-india (accessed on 09 March 2024).

Digital Infrastructure, (n.d.). Digital Infrastructure. Africa Digital Financial Inclusion Facility (adfi.org).

Digital Payments Adoption in India, (2020). *Report on Digital Payment India 2020* (npci. org.in).

Digital Payments Utsav, (n.d.). Digital Payments Utsav. Ministry of Electronics and Information Technology, Government of India (meity.gov.in).

Financial Access Survey 2021 Trends and Developments, (2021). IMF Releases the 2021 Financial Access Survey Results.

Financial Literacy and Digital Financial Inclusion: Supervisory Policy and Practice, (2022). Retrieved from: https://digitalfrontiersinstitute.org/resource/financial-literacy-and-digital-financial-inclusion-supervisory-policy-and-practice/ (torontocentre.org).

Goyal, R., Roongta, P.,Mandhata, V., Gandhi, T., Jha, G., & Lewis, N., (2022). *Digital Payments in India: A US$10 Trillion Opportunity*. https://www.bcg.com/publications/2022/future-of-digital-payments-in-india (accessed on 09 March 2024).

Grohmann, A., & Menkhoff, L., (2020). *The Relationship Between Financial Literacy and Financial Inclusion.* DIW Discussion Papers, No. 1914, Deutsches Institutfür Wirtschaftsforschung (DIW), Berlin.

HSBC Report on Indian Economy, (2018). https://assets.kpmg.com/content/dam/kpmg/in/pdf/2020/08/impacting-digital-payments-in-india.pdf (accessed on 09 March 2024).

Impact of COVID-19 on Digital Payments in India, (2020). Impact of COVID-19 on digital payments in India (assets.kpmg).

Impact of the COVID-19 Outbreak on Digital Payments, (n.d.). Digital Payments Rise in COVID-19 Time – Bankbazaar.

Incentives and Promotional Schemes, (n.d.) Incentive/Promotional Schemes, Ministry of Electronics and Information Technology, Government of India (meity.gov.in).

Lauer, K., & Lyman, T., (2015). *Digital Financial Inclusion: Implications for Customers, Regulators, Supervisors, and Standard-Setting Bodies*. Consultative Group to Assist the Poor (CGAP), February.

Lusardi, A., (2012). *Numeracy, Financial Literacy, and Financial Decision Making*. Working Paper 17821, http://www.nber.org/papers/w17821 (accessed on 09 March 2024). JEL No. D91

Meeker, M., (2017). *Internet Trends 2017 Internet Trends 2017.* Kleiner Perkins, Make History.

MeitY Concludes Digital Payment Utsav with a Challenge to Bankers, (2021). https://www.business-standard.com/article/markets/meity-concludes-digital-payment-utsav-with-a-challenge-to-bankers-121120500574_1.html (accessed on 09 March 2024).

MeitY, (2022). India made 7,422 CR digital payments in FY22 at 33% growth rate. India made 7,422 Cr. digital payments in FY22 at 33% growth rate: MeitY, Mint (livemint.com).

Modes of Digital Payment, (n.d.). Modes of Digital Payments, Ministry of Electronics and Information Technology, Government of India (meity.gov.in).

Modi, D., (2021). *Success of Digital Financial Services Rests on Rural India's Shift from Cash to Digital Payments Economy*. Retrieved from: 'Success of digital financial services rests on rural India's shift from cash to digital payments economy.' The Financial Express.

Morgan, P., Huang, B., & Trinh, L. Q., (2019). *The Need to Promote Digital Financial Literacy.* https://t20japan.org/policy-brief-need-promote-digital-financial-literacy/.

National Strategy for Financial Inclusion (NSFI): 2019–2024, (2020). Reserve Bank of India – Reports (rbi.org.in).

Nedungadi, P. P., Menon, R., Gutjahr, G., Erickson, L., & Raman, R., (2018). "Towards an inclusive digital literacy framework for digital India. *Education + Training, 60*(6), 516–528. https://doi.org/10.1108/ET-03-2018-0061.

Now Indians Will be Able to Receive Money from Overseas Via UPI, (2022). https://www.techcircle.in/2022/01/27/now-indians-will-be-able-to-receive-money-from-overseas-via-upi (accessed on 09 March 2024).

OECD, (2019). *Draft OECD Recommendation on Financial Literacy and Education.* Paris: OECD. https://www.oecd.org/daf/fin/financial-education/Recommendation-on-FL-and-FE.pdf (accessed on 09 March 2024).

Pratik, B., (2017). *Demonetization Effect: Digital Payments India's New Currency; Debit Card Transactions Surge to Over 1 Billion.* http://economictimes.indiatimes.com/industry/banking/finance/banking/digital-payments-indias-new-currency-debit-card-transactions-surge-to-over-1-billion/articleshow/58863652.cms (accessed on 09 March 2024).

Rathore, H. S., (2016). Adoption of digital wallet by consumers. *BVIMSR's Journal of Management Research, 8*, 69.

Refera, M. K., Dhaliwal, N. K., & Kaur, J., (2016). Financial literacy for developing countries in Africa. *Journal of African Studies & Development, 8*(1), 1–12. doi: 10.5897/JASD2015.0331.

Reserve Bank of India Announces Digital Payments Index for September 2021, (2022). https://www.rbi.org.in/scripts/BS_PressReleaseDisplay.aspx?prid=53123 (accessed on 09 March 2024).

Singhal, R., & Gupta, A., (2021). Impact of COVID-19 on digital payment services at towns and villages. *International Journal of Creative Research Thoughts (IJCRT) 2021*, Available at SSRN: https://ssrn.com/abstract=3947798 (accessed on 09 March 2024).

Soon, You Can Pay Via UPI All Over the World, (2019). https://economictimes.indiatimes.com/industry/banking/finance/soon-you-can-pay-via-upi-all-over-the-world/articleshow/71731402.cms?utm_source=contentofinterest&utm_medium=text&utm_campaign=cppst (accessed on 09 March 2024).

Srivastava, J., & Raghubir, P., (2008). Monopoly money: The effect of payment coupling and form on spending behavior. *Journal of Experiental Psycology Applied, 14*, 213–225.

Subha, M., & Priya, P., (2014). The emerging role of financial literacy financial planning. *IJISET – International Journal of Innovative Science, Engineering & Technology, 1*(5), 400–408.

Telecom Statistics India, (2020). *Telecom Statistics India-2020* | Department of Telecommunications | Ministry of Communication | Government of India (dot.gov.in).

TerraPay, (n.d.). *The Global Payments Highway.* https://www.terrapay.com/ (accessed on 09 March 2024).

The Indian Telecom Services Performance Indicator Report, (2022). *Performance Indicators* Reports. Telecom Regulatory Authority of India (trai.gov.in).

Toronto Centre, (2019). *Removing the Barriers to Women's Financial Inclusion.* Removing the Barriers To Women's Financial Inclusion (www.torontocentre.org).

TRAI Report Press Release, (2022). Press Release No (trai.gov.in).

Watal, R., (2017). *Digital Payments Surge 55% in 2016–17.* http://www.livemint.com/Industry/hF8D3D6bWBie6IoJzWtdZO/Digital-payments-surge-55-in-FY17-Niti-Aayog.html (accessed on 09 March 2024).

World Bank, (2018). *Financial Literacy Survey Questionnaire*. Washington, DC: World Bank.

Worldline 'India Digital Payments Report' Q1, (2022). https://www.expresscomputer.in/security/worldline-india-digital-payments-report-q1-2022/87490/ (accessed on 09 March 2024).

CHAPTER 9

Innovative Measures and Sustainable Growth of SMEs: A Special Reference to Some Selected Enterprises in Howrah

SUMEDHA MAJUMDER[1] and SHELLY DE (PANDIT)[2]

[1]*Assistant Professor, Department of Commerce, Shree Agrasain College, Howrah, West Bengal, India*

[2]*Associate Professor, Department of Commerce, Bijoy Krishna Girls' College, Howrah, West Bengal, India*

ABSTRACT

The COVID-19 pandemic lockdown had an adverse impact on the SMEs sector, resulting in negative growth in terms of production process, export-import, employment, market demand, and liquidity. The lathe industry-based SMEs in Howrah district, West Bengal, were not an exception to this crisis. Therefore, the objectives of this study are to understand, analyze, and compare the sales, purchase, and profit of selected lathe industry-based enterprises in Howrah district, West Bengal, before and after the COVID-19 pandemic. The study will also identify and analyze the survival strategies and innovative measures taken by these selected SMEs in the new normal. The study is based on both primary and secondary data. The study found that more than 50% of the enterprises experienced a decline in sales with a negative CAGR from 2018–2019 to 2020–2021. The purchase of 50% of the selected enterprises also decreased between 2018 and 2019 & 2020 and 2021. Profit also decreased for these SMEs, although some of them showed positive growth in sales and profit with a positive and high CAGR. Innovative measures such as a combination of online-offline sales, cashless

Contemporary Digital Transformation and Organizational Effectiveness in Business 4.0.
Nilanjan Ray, Sudin Bag, & Tapas Kumar Chatterjee (Eds.)

payments, home deliveries, and well-developed customer relationships played a crucial role in the existence and sustainable growth of these SMEs during the lockdown.

9.1 INTRODUCTION

In the realm of business sustainability, self-motivation emerges as the pivotal factor for achieving long-term success. This fundamental strategy has proven its validity during and after the prolonged and unforeseen lockdown caused by the COVID-19 pandemic. The global impact of the COVID-19 lockdown can be likened to a nightmarish scenario, given its devastating effects on economic activities. The outbreak of COVID-19 was first detected in India during the initial week of March 2020, leading to an unplanned lockdown imposed by the Government of India (GOI) on March 24, 2020 (https://economictimes.indiatimes.com/). The adverse consequences of the COVID-19 lockdown manifested in negative growth within the production process, a decline in export-import activities, workforce reductions, a contraction in market demand, disruptions in the supply chain, and a liquidity crisis. In March 2020, merchandise exports and imports experienced significant declines of 34.6% and 28.7% respectively, compared to the previous month of the same year (Monthly Income Report, Ministry of Finance. Furthermore, as of December 20, 2021, India recorded an unemployment rate of 8% (Statista, 2022).

One thing we have realized is that we need to adapt to the new post-COVID-19 situation, which has been referred to as the "new normal" (Hodes, 2020). Over the past two years, we have witnessed an unprecedented transformation in our lives. Business operations have shifted from offline to online mode. The forced digitalization is a direct result of the COVID-19 crisis (Amoah et al., 2021). In this vulnerable economic situation, many businesses are striving to change their business models in order to survive in the market and retain their customers. The GOI has implemented several measures to revive the economy, although not all businesses have been able to take advantage of them. On the other hand, a significant number of small and medium enterprises (SMEs) have successfully continued their business operations by embracing a positive attitude during the new normal.

The SME sector has played a crucial role in boosting the Indian economy for several years. However, the COVID-19 pandemic has significantly disrupted the normal performance of this sector. As a result, 95% of small sector enterprises have been adversely affected. Due to the unavailability of sufficient capital, their productivity has declined (Tripathi, 2021).

Howrah, the twin city of Kolkata, is popularly known as the Sheffield of the East (mapsofindia.com) due to the presence of the engineering industry in Liluah, Belur, Bally, Santragachi, Shibpur, Uluberia, Ramrajatala, and Bantra. These clusters of small engineering enterprises are operated by local people.

Like other SMEs in the country, the lathe industry-based SMEs in Howrah, West Bengal, have also faced significant challenges in running their businesses during different phases of the COVID-19 pandemic. However, a group of enterprises has made efforts to shift towards digitalization in the new normal for their sustainability. Some businesses have identified new needs and shifted their focus towards toiletries, sanitizers, masks, PPE kits, etc. Additionally, many businesses have started prioritizing customer relationship management (CRM) by offering free home delivery and operating online ordering systems, among other measures.

9.2 LITERATURE REVIEW

Adam et al. (2021) revealed that the innovation practices of SMEs have a remarkable impact on their business performance and ensure their sustainability. Thukral (2021) found that the COVID-19 pandemic has led to behavioral changes, significantly affecting entrepreneurship, especially among SMEs. SMEs are striving to overcome these challenges posed by COVID-19 through their creative and innovative approaches. Kulkarni (2020) concluded that there is a high demand for SMEs to adopt digital business operations in order to enhance their competitiveness and maintain their position in the development of the Indian economy. This study also highlighted the opportunities for SMEs to embrace digitalization through existing schemes such as Make in India, Digital India, E-governance, etc. Arora (2019) suggested a strong correlation between firm size and the adoption of digitalization. The study also identified various challenges, such as lack of appropriate technology and high costs, associated with implementing digitalization in business. Bouwman et al. (2019) commented that digital transformation has the potential to reshape the business model of SMEs. Ng et al. (2019) concluded that the financial performance of SMEs is influenced by four key factors: entrepreneurship, leadership, expertise, and innovation. To achieve financial performance, owner-managed SMEs always strive to focus on innovative products, processes, and behavior. Falahat et al. (2018) revealed the positive impact of perceived non-financial factors, perceived business growth, and perceived performance on the sustainability of businesses. Paul (2017) found that the main influencing factors for SMEs in India to choose content marketing include relative advantage, employee knowledge, level of innovativeness, and competitive pressures. Martin et al. (2014) concluded that innovation is considered the most important factor for the growth and sustainability of businesses. Rosli et al. (2013) also concluded that innovation is considered the most critical factor in today's entrepreneurial activities of SMEs, as it has a positive impact on firm performance.

9.3 RESEARCH GAP

Based on the literature review above, it has been found that no study has been conducted thus far to analyze the survival strategies of lathe industry-based SMEs in Howrah district in order to combat the economic challenges in the new normal. As a result of the COVID-19 crisis, a significant number of SMEs were forced to cease their business operations. Primary data indicates that only a few enterprises within the SMEs of Howrah district, West Bengal were able to generate profits during this challenging economic situation. Therefore, the research problem is to identify how these enterprises are successfully combating the COVID-19 pandemic and the various types of creative and innovative measures they have taken for strategic improvement and sustainability.

9.4 RESEARCH QUESTION

1. Are all of the lathe industry-based SMEs in Howrah district, which were operational before the pandemic, still in operation?
2. How did the sales, purchases, and profits of the selected enterprises in Howrah district, West Bengal, get affected before and after the COVID-19 pandemic?
3. What survival strategies and innovative measures have the selected SMEs in Howrah district adopted to survive during the COVID-19 pandemic?
4. To what extent did the various COVID incentive schemes provided by the Government offer relief to these SMEs from financial distress?

9.5 OBJECTIVE

1. The objective of this study is to comprehend, analyze, and compare the sales, purchase, and profit of the chosen enterprises in Howrah district, West Bengal, both before and after the COVID-19 pandemic.

2. Another aim is to identify and analyze the survival strategies and innovative measures implemented by these selected SMEs in the new normal.

9.6 METHODOLOGY

1. **Type of Research:** The study in analytical as well as descriptive in nature.
2. **Sources of Data:** The study is based on Primary as well as Secondary data.
   i. **Primary Data:** It is collected from a structured questionnaire administered to the owners of respective SMEs in Howrah district, West Bengal. The questionnaire contains questions related to the sales, purchases, profits, and innovative business strategies implemented by the selected enterprises for the three years: 2018–2019, 2019–2020, and 2020–2021.
   ii. **Secondary Data:** It is collected from journals, articles, books, government orders, and relevant electronic media.
3. **A Sample Size of 30:** Small-scale enterprises in Howrah, West Bengal, all of which belong to the lathe industry.
4. **Sampling Method:** Convenient sampling method is used to select the SMEs.
5. **Data Analysis:** Both statistical and logical derivations are taken into consideration. The data is analyzed using metrics such as annual growth rate (AGR), compounded annual growth rate (CAGR), standard deviation, and correlation.

9.7 ANALYSIS AND FINDINGS

9.7.1 CHANGE IN SALES FROM 2018–2019 TO 2020–2021

From Table 9.1, it was found that nearly 54% of enterprises experienced a negative AGR in their sales during the COVID-19 identified year (2020–2021). Only a few (6.67%) of them (specifically 2 enterprises) were able to increase their sales (AGR) by more than 100% during the COVID-19 identified year. Prior to the COVID-19 pandemic, in 2018–2019, the average sales of the selected enterprises were Rs. 6,926,540, which decreased to Rs. 6,359,524 with an AGR of – 8.19 in 2019–2020, and then further declined to Rs. 5,713,501 with an AGR of – 10.16 in 2020–2021. Most of the enterprises faced a decline in sales due to the nationwide prolonged lockdown for the COVID-19 pandemic.

TABLE 9.1 Sales of Selected Enterprises from 2018–2019 to 2020–2021

Sl. No.	Name of the Enterprises*	Total Sales (Rs.)					CAGR (%)
		2018–2019	2019–2020	AGR (%)	2020–2021	AGR (%)	
1.	AE	11,17,464	13,50,328	20.84	14,59,929	8.12	30.65
2.	BE	33,50,734	36,04,696	7.58	29,88,666	(17.09)	(10.81)
3.	CE	55,10,126	47,01,681	(14.67)	24,10,753	(48.73)	(56.25)
4.	DE	14,02,196	13,25,118	(5.50)	15,58,565	17.62	11.15
5.	EE	50,13,098	41,04,013	(18.13)	19,79,959	(51.76)	(60.50)
6.	FE	11,35,882	10,04,430	(11.57)	20,22,209	101.33	78.03
7.	GE	23,38,856	26,80,305	14.60	32,25,790	20.35	37.93
8.	HE	6,19,558	5,16,000	(16.71)	6,44,110	24.83	3.97
9.	IE	86,20,880	87,48,394	1.48	1,63,16,441	86.51	89.27
10.	JE	16,99,634	16,71,930	(1.63)	10,07,319	(39.75)	(40.73)
11.	KE	1,09,39,244	1,00,07,912	(8.51)	83,01,702	(17.05)	(24.11)
12.	LE	13,56,040	16,07,400	18.54	40,66,371	152.98	199.87
13.	ME	64,06,244	64,05,453	(0.012)	52,89,777	(17.42)	(17.43)
14.	NE	7,13,266	6,61,471	(7.26)	13,96,933	111.19	95.85
15.	OE	2,07,89,606	1,79,24,927	(13.78)	96,38,650	(46.23)	(53.64)
16.	PE	1,44,74,346	1,19,21,772	(17.64)	88,22,234	(26)	(39.05)
17.	QE	4,26,596	4,69,223	10	7,03,954	50.03	39.40
18.	RE	15,79,552	12,95,332	(18)	9,97,581	(23)	(36.84)
19.	SE	56,59,436	49,19,17	(13.08)	44,11,230	(10.32)	(22.06)
20.	TE	70,04,176	57,01,226	(18.60)	25,27,517	(55.67)	(63.91)
21.	UE	29,14,880	25,28,705	(13.25)	40,73,993	61.11	39.77
22.	VE	57,03,024	54,68,286	(4.12)	30,95,643	(43.39)	(45.72)
23.	WE	92,59,718	80,91,497	(12.62)	70,62,451	(12.72)	(23.73)
24.	XE	30,42,758	29,61,262	(2.68)	39,74,163	34.21	30.61
25.	YE	21,88,840	19,90,500	(9.06)	22,71,045	14.09	3.76
26.	ZE	7,58,738	11,59,425	52.81	21,62,096	86.48	184.96
27.	AAE	29,94,092	29,93,342	(0.025)	36,49,025	21.90	21.87
28.	ABE	7,03,99,468	6,57,59,904	(6.59)	6,11,56,711	(7)	(13.13)
29.	ACE	17,60,738	16,38,569	(6.94)	6,62,081	(59.59)	(62.40)
30.	ADE	86,17,014	75,73,523	(12.11)	35,28,129	(53.41)	(59.02)
	Average	69,26,540	63,59,524	(8.19)	57,13,501	(10.16)	(17.51)
	Standard Deviation	1,26,32,370	1,16,98,230	–	1,08,12,867	–	–

*To maintain privacy, original names of the enterprises are changed.

Source: Primary data.

It was observed that over 50% of the enterprises suffered a significant decrease in sales with a CAGR of – 17.51 from 2018–2019 to 2020–2021, which is highly noteworthy. During the lockdown, the supply chain mechanism was disrupted due to transport restrictions, resulting in a complete halt of business activities for the initial three months, from March 2020 to June 2020. According to the aforementioned table, the average sales of the selected enterprises decreased by 10% in 2020–2021 compared to the previous year, 2019–2020. Furthermore, the table also revealed a significant standard deviation in the sales of the selected enterprises, amounting to Rs. 12,632,370, Rs. 11,698,230, and Rs. 10,812,867 in 2018–2019, 2019–2020, and 2020–2021, respectively. However, it was also discovered that the sales gap among the selected enterprises narrowed due to the COVID-19 pandemic.

The majority of these businesses faced challenges such as a liquidity crisis, a decline in demand, order cancellations, and the fixed financial burden of employee salaries, wages, and interest expenses on borrowed funds. The study revealed that a few businesses were able to generate a significant volume of sales by implementing innovative strategies and measures, establishing themselves as examples of sustainable SMEs during this challenging period. The highest CAGR recorded was 199.87%.

9.7.2 CHANGE IN PURCHASES FROM 2018–2019 TO 2020–2021

According to Table 9.2, we discovered that the purchase of 50% of the selected enterprises decreased between 2018 and 2019 & 2020 and 2021. The other 50% of the enterprises successfully ran their businesses, resulting in a positive change in purchases to meet the market demand for their merchandise. Moreover, the average purchases of the selected enterprises increased from Rs. 48,33,104 to Rs. 61,20,455 and then to Rs. 67,67,422 in the years 2018–2019, 2019–2020, and 2020–2021, with an AGR of 26.84% and 10.57% for the years 2019–2020 and 2021–2022, respectively. The average CAGR of these enterprises stood at 40.25%. This positive result in purchases reveals that the input costs have also increased. The average purchases of one selected enterprise increased by 283.07% from 2018–2019 to 2020–2021. Another thing to note is that the standard deviation in purchases of 30 enterprises increased from 2018–2019 to 2019–2020 and from 2019–2020 to 2020–2021, indicating that the operations of all of them were not adequate to combat COVID-19.

TABLE 9.2 Purchases of Selected Enterprises from 2018–2019 to 2020–2021

Sl. No.	Name of the Enterprises*	Total Purchases (Rs.)					CAGR (%)
		2018–2019	2019–2020	AGR (%)	2020–2021	AGR (%)	
1.	AE	16,08,964	12,23,272	(23.97)	16,79,348	37.28	4.37
2.	BE	28,55,908	29,88,812	4.65	21,01,617	(29.68)	(26.41)
3.	CE	43,49,668	18,26,634	(58)	20,74,134	13.55	(52.32)
4.	DE	8,29,568	8,36,592	0.85	13,48,480	61.19	62.55
5.	EE	41,51,502	10,38,438	(74.99)	20,97,283	101.97	(49.48)
6.	FE	10,44,578	10,04,430	(3.84)	13,68,772	36.27	31.04
7.	GE	23,47,926	26,80,305	14.16	19,76,628	(26.25)	(15.81)
8.	HE	23,47,926	5,16,000	(78.02)	2,79,113	(45.91)	(88.11)
9.	IE	65,89,256	87,48,394	32.77	1,17,86,917	34.73	78.88
10.	JE	3,16,400	16,71,930	428.42	4,58,177	(72.60)	44.81
11.	KE	1,03,04,622	1,00,07,912	(2.88)	71,90,577	(28.15)	(30.22)
12.	LE	41,04,380	16,07,400	(60.84)	34,63,689	115.48	(15.61)
13.	ME	63,97,938	64,05,453	0.12	30,22,376	(52.82)	(52.76)
14.	NE	6,53,564	6,61,471	1.21	9,15,846	38.46	40.13
15.	OE	1,63,30,620	1,79,24,927	9.76	1,14,19,997	(36.29)	(30.07)
16.	PE	88,25,440	1,19,21,772	35.08	60,83,669	(48.97)	(31.07)
17.	QE	2,86,350	4,69,223	63.86	3,02,348	(35.56)	5.59
18.	RE	2,51,444	12,95,332	415.16	4,73,869	(63.42)	88.46
19.	SE	54,51,868	49,19,107	(9.77)	44,08,668	(10.38)	(19.13)
20.	TE	63,93,762	57,01,226	(10.83)	23,92,122	(58.04)	(62.59)
21.	UE	32,07,048	25,28,705	(21.15)	38,87,992	53.75	21.23
22.	VE	47,13,058	54,68,286	16.02	23,48,097	(57.06)	(50.17)
23.	WE	50,78,344	80,91,497	59.33	37,87,992	(53.19)	(25.41)
24.	XE	28,60,386	29,61,262	3.53	14,45,578	(51.18)	(49.46)
25.	YE	1,47,354	19,90,500	1,250	3,78,426	(80.99)	156.81
26.	ZE	8,88,007	11,59,425	30.56	17,51,880	51	97.28
27.	AAE	16,88,760	29,93,342	77.25	16,93,787	(43.41)	0.30
28.	ABE	3,60,06,226	6,57,59,904	82.63	12,03,40,624	83	234
29.	ACE	1,29,157	16,38,569	1,169	4,94,763	(69.81)	283.07
30.	ADE	45,97,874	75,73,523	64.72	20,49,905	(72.93)	(55.42)
	Average	48,25,263	61,20,455	26.84	67,67,422	10.57	40.25
	Standard Deviation	67,63,093	1,17,68,584	–	2,12,797,678	–	–

*In order to maintain privacy, the original names of the companies have been changed.

Source: Primary data.

9.7.3 TREND OF PROFIT FROM 2018–2019 TO 2020–2021

From Table 9.3, it is observed that the average profit is decreasing in 2020–2021 compared to 2018–2019 for the selected enterprises. Around 70% of the selected enterprises, which is 27 enterprises, experienced a decline in their earning capacity in 2019–2020 compared to 2018–2019 due to the unprecedented nationwide lockdown. Additionally, the average profit of these enterprises also fell by 7% in 2019–2020 compared to 2018–2019. Most of the enterprises saw a reduction in profit of more than 10% in 2019–2020. The situation became even more alarming in 2020–2021 due to the lengthy unplanned lockdown. The average profit of the selected enterprises fell by 10.97% in 2020–2021 compared to 2019–2020. The average Compound Annual Growth Rate (CAGR) stood at – 17.20. The standard deviation in net profit among the selected enterprises decreased from 2018–2019 to 2020–2021.

TABLE 9.3 Net Profit of Selected Enterprises from 2018–2019 to 2020–2021

Sl. No.	Name of the Enterprises*	Net Profit (Rs.)					CAGR (%)
		2018–2019	2019–2020	AGR (%)	2020–2021	AGR (%)	
1.	AE	90,515	1,08,026	19.35	1,19,714	10.82	32.26
2.	BE	2,74,760	2,91,980	6.27	2,39,093	(18.11)	(12.98)
3.	CE	4,46,320	3,85,537	(13.62)	1,92,860	(49.98)	(56.79)
4.	DE	1,13,578	1,08,660	(4.33)	1,24,685	14.75	9.78
5.	EE	4,06,127	3,36,529	(17.14)	1,58,397	(52.93)	(61)
6.	FE	90,870	82,363	(9.36)	1,61,777	96.42	78.03
7.	GE	1,87,108	2,17,105	16.03	2,64,515	21.84	41.37
8.	HE	50,804	41,280	(18.75)	52,173	26.39	2.69
9.	IE	6,89,670	7,17,368	4.02	13,21,632	84.23	91.63
10.	JE	1,39,370	1,35,426	(2.83)	80,586	(40.49)	(42.18)
11.	KE	8,86,079	8,00,633	(9.64)	6,64,136	(17.05)	(25.01)
12.	LE	1,08,483	1,31,807	21.50	3,29,376	149.89	67.06
13.	ME	5,18,906	5,25,247	1.22	4,23,182	(19.43)	(18.45)
14.	NE	57,061	54,241	(4.94)	1,11,755	106.03	95.85
15.	OE	16,83,958	14,51,919	(13.78)	7,71,092	(46.89)	(54.21)
16.	PE	11,57,948	9,65,664	(16.61)	7,23,423	(25.09)	(37.53)
17.	QE	34,554	38,476	11.35	57,020	48.19	65
18.	RE	1,27,943	1,06,217	(16.98)	79,806	(24.87)	(37.62)
19.	SE	4,52,755	4,03,367	(10.91)	3,57,310	(11.42)	(21.08)
20.	TE	5,74,342	4,61,799	(19.60)	2,02,201	(56.21)	(64.79)
21.	UE	2,39,020	2,04,825	(14.31)	3,25,919	59.12	36.36

TABLE 9.3 *(Continued)*

Sl. No.	Name of the Enterprises*	Net Profit (Rs.)					CAGR (%)
		2018–2019	2019–2020	AGR (%)	2020–2021	AGR (%)	
22.	VE	4,56,242	4,48,399	(1.72)	2,50,747	(44.08)	(45.04)
23.	WE	7,40,777	6,63,503	(10.43)	5,72,058	(13.78)	(22.78)
24.	XE	2,43,421	2,39,862	(1.46)	3,25,881	35.86	33.88
25.	YE	1,75,107	1,61,231	(7.92)	1,86,226	15.50	6.35
26.	ZE	62,217	93,913	50.94	1,72,968	84.18	178
27.	AAE	2,39,527	2,45,454	2.47	2,95,571	20.42	23.40
28.	ABE	56,31,957	53,92,312	(4.26)	49,53,694	(8.13)	(12.04)
29.	ACE	1,44,380	1,32,724	(8.07)	52,966	(60.09)	(63.31)
30.	ADE	7,06,595	6,13,455	(13.18)	2,82,250	(53.99)	(60.05)
	Average	5,57,680	5,18,644	(7)	4,61,767	(10.97)	(17.20)
	Standard Deviation	10,28,459	9,74,906	–	8,90,909	–	–

*To ensure confidentiality, the names of the original companies have been altered.

Source: Primary data.

9.7.4 CHANGE IN PROFIT AND CHANGE IN SALES ANALYSIS

Table 9.4 reveals that for most of the selected enterprises, the profit decreased along with the decrease in sales over the past three years. The direction of change in profit and change in sales shows a strong positive correlation, with coefficients of correlation of 1 and 0.998554481, respectively. This indicates a clear relationship between sales and profit for the selected enterprises. The likely explanation for this could be the implementation of innovative measures by the SMEs.

TABLE 9.4 Change in Sales and Change in Profit*

Sl. No.	Name of the Enterprises	Absolute Change in Sales from 2018–2019 to 2019–2020 (Rs.)	Absolute Change in Net Profit from 2018–2019 to 2019–2020 (Rs.)	Absolute Change in Sales from 2019–2020 to 2020–2021 (Rs.)	Absolute Change in Net Profit from 2019–2020 to 2020–2021 (Rs.)
1.	AE	2,32,864	17,511	1,09,601	11,688
2.	BE	2,53,962	17,220	–6,16,030	–52,887
3.	CE	–8,08,445	–60,783	–22,90,928	–1,92,677
4.	DE	–77,078	–4,918	2,33,447	16,025
5.	EE	–9,09,085	–69,598	–21,24,054	–1,78,132
6.	FE	–1,31,452	–8,507	10,17,779	79,414

TABLE 9.4 *(Continued)*

Sl. No.	Name of the Enterprises	Absolute Change in Sales from 2018–2019 to 2019–2020 (Rs.)	Absolute Change in Net Profit from 2018–2019 to 2019–2020 (Rs.)	Absolute Change in Sales from 2019–2020 to 2020–2021 (Rs.)	Absolute Change in Net Profit from 2019–2020 to 2020–2021 (Rs.)
7.	GE	3,41,449	29,997	5,45,485	47,410
8.	HE	–1,03,558	–9,524	1,28,110	10,893
9.	IE	1,27,514	27,698	75,68,047	6,04,264
10.	JE	–27,704	–3,944	–6,64,611	–54,840
11.	KE	–9,31,332	–85,446	–17,06,210	–1,36,497
12.	LE	2,51,360	23,324	24,58,971	1,97,569
13.	ME	–791	6,341	–11,15,676	–1,02,065
14.	NE	–51,795	–2,820	7,35,462	57,514
15.	OE	–28,64,679	–2,32,039	–82,86,277	–6,80,827
16.	PE	–25,52,574	–1,92,284	–30,99,538	–2,42,241
17.	QE	42,627	3,922	2,34,731	18,544
18.	RE	–2,84,220	–21,726	–2,97,751	–26,411
19.	SE	–7,40,329	–49,388	–5,07,877	–46,057
20.	TE	–13,02,950	–1,12,543	–31,73,709	–2,59,598
21.	UE	–3,86,175	–34,195	15,45,288	1,21,094
22.	VE	–2,34,738	–7,843	–23,72,643	–1,97,652
23.	WE	–11,68,221	–77,274	–10,29,046	–91,445
24.	XE	–81,496	–3,559	10,12,901	86,019
25.	YE	–1,98,340	–13,876	2,80,545	24,995
26.	ZE	4,00,687	31,696	10,02,671	79,055
27.	AAE	–750	5,927	6,55,683	50,117
28.	ABE	–46,39,564	–2,39,645	–46,03,193	–4,38,618
29.	ACE	–1,22,169	–11,656	–9,76,488	–79,758
30.	ADE	–10,43,491	–93,140	–40,45,394	–3,31,205
	Average	–5,67,016	–39,036	–6,46,023	–56,877
	Correlation	1		0.998554481	

*In order to maintain confidentiality, the names of the original companies have been modified.

Source: Primary data.

Figures 9.1 and 9.2 present the graphical representation of the absolute changes in sales and profits of the selected SMEs in Howrah, West Bengal from 2018–2019 to 2019–2020 and from 2019–2020 to 2020–2021.

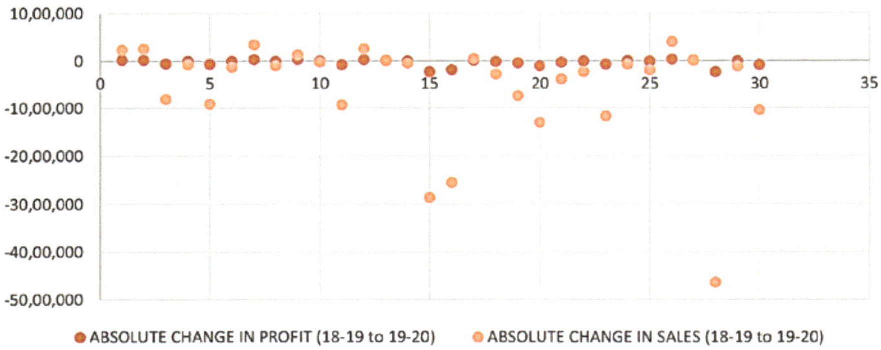

FIGURE 9.1 Absolute changes in sales and profits from 2018–2019 to 2019–2020.
Note: To uphold confidentiality, the names of the original companies have been altered.
Source: Primary data.

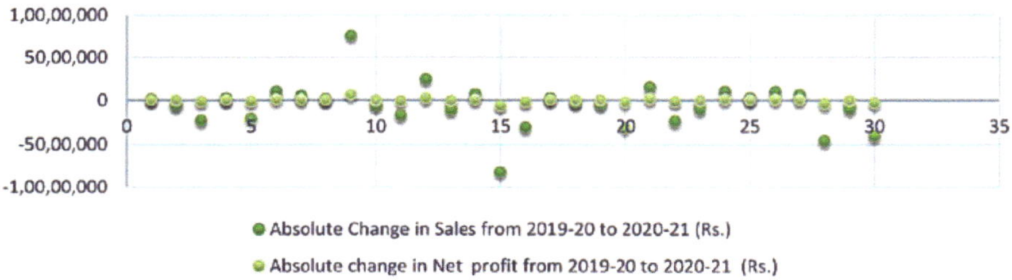

FIGURE 9.2 Absolute changes in sales and profits from 2019–2020 to 2020–2021.
Note: To keep things confidential, we changed the names of the original companies.
Source: Primary data.

9.7.5 INNOVATIVE MEASURES TAKEN BY THE SMES

According to our study, the changes in business strategies have been the primary means for selected SMEs to survive the COVID-19 pandemic. The adoption of digitalization in payment systems and marketing through social media has also been a significant innovative strategy. The main objective has been to enhance customer relations, retain existing customers, and increase market share to gain a competitive advantage. Additionally, these businesses have sought external financial support from various financial institutions and banks to ensure smooth operations.

To mitigate the decline in demand and revenue, small businesses have implemented a range of new initiatives. The prolonged lockdown due to COVID-19 has compelled them to shift operations online, including remote work for employees, home delivery of goods, and digital payment transactions. Establishing partnerships with e-commerce platforms has enabled these SMEs to boost sales and overcome challenges in selling products in the global market. The combination of online sales, cashless payments, and home deliveries has played a crucial role in sustaining these SMEs during the lockdown. Our key findings are as follows:

1. **Changed Business Model:** Many SMEs are transitioning to the e-business model, which can reduce operating costs by implementing a 'work from anywhere' strategy.
2. **Integration of Payment Methods:** Both traditional and digital payment methods are being used by businesses to sustain themselves during these challenging times. In addition to online banking, several UPI apps such as PhonePe, GPay, Paytm, etc., are being utilized to facilitate payment transactions.
3. **Increased Customized Services:** In order to retain customers, businesses are striving to provide more personalized services. They are accepting online orders and delivering merchandise through their own delivery services.
4. **E-Supply Chain Mechanism:** A majority of SMEs are adopting an E-supply chain mechanism, following the just-in-time method of stock management to reduce inventory costs.
5. **Sale of 'COVID Goods':** To overcome revenue shortfalls, a significant number of SMEs have started producing 'COVID goods' such as hand sanitizers, surface disinfectants, face masks, PPE kits, gloves, hair cover caps, etc.
6. **Adoption of New Pricing Strategy:** Our study indicates that SMEs are implementing price escalations on their merchandise to compensate for the revenue shortfall caused by low demand.
7. **The Emergency Credit Line Guarantee Scheme (ECLGS):** This scheme was launched by the GOI to provide 100% guarantee coverage to banks and NBFCs, enabling them to extend emergency credit facilities to MSMEs to meet their additional working capital requirements or provide additional loans in view of the COVID crisis (www.eclgs.com). Most of the respondents in our study mentioned that they have benefited from these additional funds, as loans were allocated by nationalized banks to combat the COVID-19 situation.

8. **Atmanirbhar Bharat Abhiyan:** The GOI announced a financial aid package worth Rs. 20 lakh crores through the Atmanirbhar Bharat Abhiyan to alleviate economic hardships during the COVID-19 pandemic (aatmanirbharbharat.mygov.in). The central government updated the definition of MSMEs in order to extend the benefits of these financial packages. Specifically, small and medium firms were provided with a collateral-free automatic loan of Rs. 3 lakh crores to aid in their recovery from the adverse effects of the COVID-19 disaster (msme.gov.in).

9.8 CONCLUSION

We observed that most of the enterprises suffered losses in the financial year 2020–2021 compared to 2019–2020. The supply chain mechanism was significantly affected during the lockdown. These enterprises faced various challenges including liquidity crisis, demand crunch, order cancellations, and the financial burden of salaries, wages, and interest expenses on borrowed money. However, a few businesses were able to generate substantial profits by implementing innovative strategies, positioning themselves as examples of sustainable SMEs during this challenging time. These SMEs focused on changing their business strategies to survive the COVID-19 pandemic. They embraced digitalization in payment systems and online marketing, which proved to be major innovative strategies. Additionally, they prioritized improving customer relations to gain a competitive advantage. Some businesses also sought external financial support from financial institutions and banks to ensure successful operations.

KEYWORDS

- **COVID-19 pandemic lockdown**
- **digitalization**
- **financial institutions**
- **Howrah**
- **innovative measures**
- **small and medium enterprises**
- **sustainable growth**

REFERENCES

Adam, N. A., & Alarifi, G., (2021). *Innovation Practices for Survival of Small and Medium Enterprises (SMEs) in the COVID-19 Times: The Role of Essential Support*. Nature public health emergency collection. https://innovation-entrepreneurship.springeropen.com/articles/10.1186/s13731-021-00156-6 (accessed on 09 March 2024).

Arora, A. K., (2019). An analysis of implementation of digitalization in SMEs in India. *International Journal of Online Marketing*. ISSN: 2156-1753; E-ISSN: 2156-1745. https://www.researchgate.net/publication/334144978_An_Analysis_of_Implementation_of_Digitalisation_in_SMEs_in_India (accessed on 09 March 2024).

Bouwman, H., Nikou, S., & Reuver, M. De. *Digitalization, Business Models and SMEs: How do Business Model Innovation Practices Improve Performance of Digitalizing SMEs?* https://www.sciencedirect.com/science/article/abs/pii/S0308596117304238 (accessed on 09 March 2024).

Falahat, M., Tehseen, S., & Horne, C. V., (2018). Entrepreneurial innovativeness and its impact on SMEs' performances. *International Journal of Entrepreneurship, 22*(3). https://www.researchgate.net/publication/326718170_Entrepreneurial_innovativeness_and_ITS_impact_on_smes%27_performances (accessed on 09 March 2024).

https://www.india.gov.in/spotlight/building-atmanirbhar-bharat-overcoming-covid-19 (accessed on 09 March 2024).

https://msme.gov.in/whatsnew/atmanirbhar-presentation-part-1-business-including-msmes-13-5-2020 (accessed on 09 March 2024).

https://www.eclgs.com/ (accessed on 09 March 2024).

https://www.mapsofindia.com/west-bengal/tourism/howrah.html (accessed on 09 March 2024).

https://www.ncbi.nlm.nih.gov/pmc/articles/PMC8437806/ (accessed on 09 March 2024).

https://www.nia.nih.gov/research/blog/2020/03/covid-19-adjusting-new-normal (accessed on 09 March 2024).

Kulkarni, M., (2020). A research paper on the need for digitalization of SMEs in India. *Proceedings of 2nd International Research e-Conference on Corporate Social Responsibility*. Organized by Dr. D. Y. Patil Vidyapeeth Pune (Deemed to be University); Global Business Schools & Research Centre, Pune-411033; ISBN: 978-93-87901-17-9. https://www.neliti.com/publications/336033/a-research-paper-on-the-need-for-digitalization-of-smes-in-india (accessed on 09 March 2024).

Martin, M. S., & Namusonge, M. J., (2014). Influence of innovation on small and medium enterprise (SME) growth — A case of garment manufacturing industries in Nakuru County. *International Journal for Innovation Education and Research*. ISSN: 2411-2933. https://r.search.yahoo.com/_ylt=Awrx.q6EDf1lSM4GGa67HAx.;_ylu=Y29sbwNzZzMEcG9zAzEEdnRpZAMEc2VjA3Ny/RV= 2/RE=1711111685/RO=10/RU=https%3a%2f%2fir-library. ku.ac.ke%2fbitstream%2fhandle%2f123456789%2f14804%2fINFLUENCE%2520OF%2520INNOVATION% 2520ON%25 20SMALL%2520AND%2520MEDIUM% 2520ENTERPRISE.pdf/RK=2/RS=hDxsxaRhOOkohezaRc3M3rcRfgU- (accessed on 09 March 2024).

Ng, H. S., Kee, D. M. H., & Ramayah, T., (2019). Examining the mediating role of innovativeness in the link between core competencies and SME performance. *Journal of Small Business and Expertise Development*. https://www.emerald.com/insight/content/doi/10.1108/JSBED-12-2018-0379/full/html (accessed on 09 March 2024).

One year since a complete lockdown was announced, we look back on how India fought COVID (2021). https://economictimes.indiatimes.com/news/india/one-year-since-a-complete-lockdown-was-announced-we-look-back-on-how-india-fought-covid/first-lockdown-announced/slideshow/81662838.cms (accessed on 09 March 2024).

Paul, S., (2017). A study on factors influencing the adoption of content marketing by small and medium enterprises (SMEs) in India. The International Journal Research Publications; ISSN: 2251-1547. *Research Journal of Commerce & Behavioral Science*. https://typeset.io/papers/a-study-on-factors-influencing-the-adoption-of-content-1zk2w1ylzx (accessed on 09 March 2024).

Rosli, M. M., & Sidek, S., (2013). The impact of innovation on the performance of small and medium manufacturing enterprises: Evidence from Malaysia. *Journal of Innovation Management in Small & Medium Enterprise, 2013*. https://www.researchgate.net/publication/314018152_The_Impact_of_Innovation_on_the_Performance_of_Small_and_Medium_Manufacturing_Enterprises_Evidence_from_Malaysia (accessed on 09 March 2024).

Thukral, E., (2021). *COVID-19: Small and Medium Enterprises Challenges and Responses with Creativity, Innovation, and Entrepreneurship*. https://onlinelibrary.wiley.com/doi/10.1002/jsc.2399 (accessed on 09 March 2024).

CHAPTER 10

Exploring Key Factors Apart from Price and Quality and Effective Ad Media to Get a Foothold in the Fast-Moving Consumer Goods (FMCG) Segment

SOUMYA MUKHERJEE,[1] MRINAL KANTI DAS,[2] and SWATI PAL[3]

[1]*Associate Professor, Department of Management Studies, Techno India (Hooghly Campus), Chinsura, Hooghly, West Bengal, India*

[2]*Assistant Professor, Department of Commerce, Kanchrapara College, Kanchrapara, West Bengal, India*

[3]*Associate Professor, Department of Management, School of Business, Adamas University, Barasat, Kolkata, West Bengal, India*

ABSTRACT

One of the fastest-growing sectors in the Indian economy is FMCG (Kavitha, 2010). Despite its growth, it has become a challenging task for marketers to establish a strong presence in the competitive industry due to the rapid changes in customers' tastes and preferences. Therefore, formulating a strategy that can make a significant impact in this competitive industry has become a daunting task. To address this, we first attempted to identify the most effective advertising media by utilizing the Thurstone Scale. This attitude measurement scale provides valuable insights into the media choices of prospects, enabling marketers to reach them at an optimal level. The study revealed that prospects still have a strong preference for audio-visual media, closely followed by digital media. Additionally, we employed both exploratory and confirmatory factor analysis to identify the key factors, other than price and quality, that have the most persuasive influence on the target audience. Our

Contemporary Digital Transformation and Organizational Effectiveness in Business 4.0.
Nilanjan Ray, Sudin Bag, & Tapas Kumar Chatterjee (Eds.)

study identified four factors - Advertisement & Celebrity Endorsement, Pack Size, Availability, and Retail & Shop Display. Understanding these factors will not only assist marketers in developing effective marketing strategies but also help them gain acceptance among prospects.

10.1 INTRODUCTION

The fast-moving consumer goods (FMCG), the fourth largest sector in our country, have a significant contribution to the growth of India's GDP. According to the report published in The Economic Times (Jan 31, 2022), India's FMCG market grew at a rate of 16%, hitting a nine-year high in 2021. It is worth noting that this growth occurred even during the pandemic. Based on current trends, the FMCG market in India is projected to grow at a CAGR of 14.9% and reach US $220 billion by 2025 (www.ibef.org/industry/fmcg).

In almost all countries, a significant portion of consumers' budget is allocated to the consumption of FMCGs (Raghunath, 2018). Despite fierce competition in both the organized and unorganized segments, the FMCG industry has excelled in designing quality distribution networks, achieving low levels of penetration, and initially operating at low costs with lower per capita consumption (Patil, 2016). However, due to changing customer preferences, survival and growth in this industry have become challenging for marketers. Therefore, marketers must formulate effective strategies to establish their presence and thrive in this competitive landscape.

Given the intense competition, marketers must carefully craft marketing strategies to establish a strong foothold in the industry. In this research work, we aim to assess the most effective media for promoting FMCG products and reaching the maximum target audience. Additionally, we will explore the factors that play a pivotal role in attracting prospects to FMCG products.

10.2 THEORETICAL BACKGROUND

The FMCG sector is considered India's fourth-largest sector, making a significant contribution to the growth of India's GDP (Kavitha, 2010). Globalization compels marketers to identify the evolving needs and desires of regular customers (Vibhuti et al., 2014), as there is a definite shift in consumer buying behavior. Therefore, it is crucial for marketers to demonstrate their value in this competitive industry. Over time, the FMCG market has become more volatile and complex due to the presence of multinational companies, domestic companies, and the unorganized sector (Patil, 2016).

Marketers must identify the factors that influence impulsive customer buying behavior (Tauseef, 2011), taking into account the growth of the retail market in India. Hence, it is high time for marketers to explore significant factors that play a pivotal role in the purchase intention of both existing and prospective consumers for FMCG products (Chatterjee et al., 2018). This provides marketers with the opportunity not only to be knowledgeable about the changing wants and needs of prospects but also to excel in the competitive world.

10.2.1 IDENTIFYING FACTORS PERSUADING CUSTOMERS OF FMCG PRODUCTS

Brand name (Bulle, 2020) and product quality (Sugrova, 2017) are the two primary factors that attract prospects towards their respective brands. Given the competitive nature of the market, it is crucial to explore other significant factors that can capture the attention of prospects towards FMCG brands. Somireddy (2018) emphasized the significant role of advertising in influencing consumer purchase decisions for selected FMCG products. He further highlighted that advertising plays a crucial role in effectively communicating the product message to the target customers. Deliya (2012) emphasized how package design and labeling of FMCG products contribute to the brand evolution of the respective products. In addition to providing usage instructions and contents, labeling provides essential information about ingredients or raw materials, usage warnings, and product care instructions. It is also true that consumers seek products that appear trustworthy, making it a key variable (Ali et al., 2012). Celebrity endorsement has emerged as one of the most effective promotional tools to entice consumers towards FMCG products (Choi & Rifon, 2012). Despite being aware of all these key factors, marketers cannot ignore the influence of price and quality in attracting customers, especially in rural areas (Narayana & Mathew, 2015). Furthermore, the availability of preferred brands in stores also significantly impacts sales volume (Rafaya et al., 2017).

10.2.2 IMPACT OF AD MEDIA TO PERSUADE CUSTOMERS OF FMCG PRODUCTS

The worth of different media, such as TV, radio, newspapers, magazines, and the Internet, is considered by consumers from various perspectives, and its long-term impact also varies considerably (Doyle & Saunders, 1990).

For example, TV can provide a comprehensive high-quality audio-visual experience, making it highly suitable for product categories that require physical demonstration. In terms of audio content, radio is the most persuasive medium for promoting businesses to local marketers (Petersen, 2018). Berkowitz et al. (2001) emphasized the impact of different media platforms as consumers' memory changes over time. Initially, TV advertisements have a definite influence on consumers' memory, but it diminishes over time. On the other hand, magazines have a low but persistent impact on consumers as they can read magazines at their own convenience. Marketers use TV and radio to reach the maximum target audience (Farooq & Latif, 2011; Fill, 2009). However, they cannot ignore print media to this day, as newspapers have the largest reach with over 330 million readers (Jayaraj, 2011). Advertisers still rely on traditional platforms such as TV and print media for advertising (Statista, 2016). The introduction of JIO has brought a revolutionary change to digitalization. In contrast to traditional media, the Internet has more control over users in terms of the content they choose to consume (Bezjian-Avery et al., 1998). The digital platform encompasses all social media platforms – Facebook, Twitter, Instagram, YouTube, email (Ducoffe, 1996; Gilaninia et al., 2013; Mishra & Vashiath, 2017; Trivedi, 2017), and its impact in the present day is noteworthy.

10.3 OBJECTIVES OF THE STUDY

Despite being one of the fastest-growing sectors, marketers of FMCG companies face a challenging task in proving their worth in the market. Therefore, it has become essential to develop an effective marketing strategy. To create a diligent marketing strategy, it is crucial for marketers to have sufficient knowledge about the factors that persuade prospects to trust the brand. In addition to identifying these factors, it is also necessary to determine the most effective advertising media to reach prospects at the optimal level. Considering the market's volatility, we have established the following specific objectives for this study:

- To explore the effective advertising media to disseminate information about the FMCG products and reach the optimum number of target audience.
- To uncover the factors that entice prospective customers towards FMCG brands.

10.4 METHODOLOGY

The present study is cross-sectional and empirical in nature, encompassing different parts of West Bengal. The study involved 440 respondents who are primarily customers of FMCG products. Respondents were selected using a simple random sampling method. The study focused on both the cosmetics and toiletries segments of FMCG products, specifically Fairness Cream, Shampoo, Deodorant, and Soap. Data was collected using a structured questionnaire with a 5-point Likert scale. This technique is valuable as it does not rely on a judgment group to generate item scale values (Edwards & Kenny, 1946; Barclay & Weaver, 1962; Price & Mueller, 1986). Factor analysis was employed to examine how principal constructs influence responses across various measured variables (DeCoster, 1998). Additionally, the Thurstone scale (Thurstone, 1929) was used to validate which media is superior for selecting FMCG brands. This scale helps measure respondent attitudes (Cormack, 2000; Bowling, 2002). Feedback was obtained from individuals of different age groups, including housewives, students, government employees, private employees, businessmen, professionals, and others. This diverse representation ensures the study's relevance and realism (Table 10.1).

TABLE 10.1 Demographic Profile of Respondents

Variables	Description	Frequency	Percentage (%)
Gender	Male	185	42.05
	Female	255	57.95
Age	21 years–30 years	95	21.59
	31 years–40 years	107	24.32
	41 years–50 years	104	23.64
	51 years–60 years	70	15.91
	61 years and above	64	14.54
Occupation	Student	88	20.00
	Government employee	54	12.27
	Private employee	81	18.41
	Housewives	137	31.14
	Professional	42	9.54
	Others	38	8.64
Income	Below Rs. 15,000	96	21.82
	Between Rs. 15,000 and Rs. 30,000	98	22.27
	Between Rs. 30,000 and Rs. 45,000	94	21.37
	Between Rs. 45,000 and Rs. 60,000	97	22.04
	Above Rs. 60,000	55	12.50

10.5 ANALYSIS AND FINDINGS

10.5.1 ASSESSING THE AD MEDIA FOR PROMOTING FMCG BRANDS WITH THE HELP OF THURSTONE SCALE

The Thurstone Scale, better known as an attitude measurement scale, is highly relevant for this study. It allows us to logically determine the attitudes of prospects towards media preferences, providing us with in-depth knowledge about FMCG products (Table 10.2).

TABLE 10.2 Observed Proportions Preferring Media X (Column of the Table) to Media Y (Row of Table)

Media Y	**Preferred Media (Media X)**			
	Print Media	**Audio-Visual Media**	**Broadcasting Media**	**Digital Media**
Print media	–	0.74	0.27	0.60
Audio-visual media	0.26	–	0.08	0.43
Broadcasting media	0.73	0.92	–	0.78
Digital media	0.40	0.57	0.22	–

Source: Primary data.

To construct the matrix mentioned above, two advertising media were selected at a time, and the respondents were asked to choose one alternative in each of the six cases. This paired comparison matrix was designed to limit the maximum number of pairs to six for this study. Finally, the responses collected through simple random sampling technique were accumulated. The observed proportions will provide us with an understanding of the distance between different media.

Under the assumption of normality, the distance between two media can be interpreted based on these proportions. These proportions are treated as probabilities, and the corresponding Z values, symbolically represented as ordinates, were obtained from a normal distribution table. Z values typically exhibit a symmetric pattern around zero. If the proportion has a value less than 0.5, the corresponding Z value will be negative. Conversely, if the proportion has a value greater than 0.5, the Z value will be positive. In Table 10.3, we have summarized the Z values.

Table 10.3 indicates that the target audience of FMCG products has a clear preference for audio-visual media. It also shows that prospects are more inclined towards digital media compared to print media. From the customers' perspective, our study suggests that broadcasting media is less popular.

TABLE 10.3 Z-Values Corresponding Top Earlier

Media Y	Preferred Media (Media X)			
	Print Media	Audio-Visual Media	Broadcasting Media	Digital Media
Print media	–	0.64	–0.61	0.25
Audio-visual media	–0.64	–	–1.41	–0.18
Broadcasting media	0.61	1.41	–	0.77
Digital media	–0.25	0.18	–0.77	–
Total	–0.28	2.23	–2.79	0.84
Mean (Z)	–0.07	0.5575	–0.6975	0.21
R* (case-V scale value)	0.6275	1.255	0	0.9075
Rank	3rd	1st	4th	2nd

Source: Authors' calculation.

10.5.2 EXTRACTING FACTORS THROUGH EXPLORATORY FACTOR ANALYSIS

To strategically explore the influencing factors, we initially employed exploratory factor analysis. The Kaiser-Meyer-Olkin (KMO) measure was exclusively used to ensure the adequacy of the sample. The value, ranging between 0.5 and 1.0, precisely indicates that we can proceed with the factor analysis in terms of the sample (Malhotra et al., 2008). In this study, the KMO measure value is 0.713, which justifies the adequacy of the sample (*see* Table 10.4). The researchers specifically utilized Bartlett's test of sphericity to assess the suitability of the data. A higher value of Bartlett's test of sphericity indicates a greater chance for the factor model to be appropriate. In our study, the approximate chi-square value is 5979.144 at 91 degrees of freedom, which implies significance at the 5% level. The findings of the KMO measure and Bartlett's Test provide the opportunity to extract the influencing factors that persuade prospects to buy FMCG products.

TABLE 10.4 KMO and Bartlett's Test

Kaiser-Meyer-Olkin Measure of Sampling Adequacy		0.713
Bartlett's Test of Sphericity	Approx. chi-square	5979.144
	df	91
	Sig.	0.000

In factor analysis, we place emphasis on eigenvalues that are greater than 1.0. We exclude other factors that have eigenvalues less than that threshold. The present study indicates that there are four factors to be considered based on eigenvalue. These four factors account for a cumulative percentage of variance of 78.932% of the total variance. This suggests that four factors are reasonable in this situation and represent a significant portion of the data.

Factor loading is a crucial aspect to consider in factor analysis. In our study, the factor loadings range between 0.725 and 0.996. To interpret the factors, the coefficients of the factor matrix also need to be carefully assessed. We applied the Maximum Likelihood Method for factor extraction (Joreskog, 1967). For rotation, we opted for Promax with Kaiser Normalization procedure (Tabachnick & Fiddell, 2007), which aims to minimize the number of variables with high loadings on a factor, thus increasing the interpretability of the factors (Table 10.5).

TABLE 10.5 Naming of the Factors

Factor No.	Name of Factors	Label	Statements	Factor Loading
1.	Retailers and shop display	S6	Shop display affects consumers' buying decision.	0.884
		S8	Visibility of FMCG brand in self-influence prospective customers.	0.856
		S13	Customers are often keen to listen to the advice of the retailers.	0.903
		S17	Sometimes retailers influence customers to buy a particular brand of FMCG.	0.924
2.	Advertisement and celebrity endorsement	S1	Advertisement helps to make choice among the available brands.	0.871
		S3	Advertisement helps to identify hidden needs with the available brand.	0.907
		S11	Celebrity endorsement makes you more passionate about the FMCG brand.	0.863
		S14	Advertisement helps to develop strong attachment with the FMCG brand.	0.899
3.	Pack size	S2	Pack sizes of FMCG brand increase the affordability of customers.	0.938
		S4	Reduced pack sizes helped to exceeding sales of traditional pack sizes.	0.865
		S10	FMCG companies have come up with reduced pack size to boost consumption of consumers.	0.914

TABLE 10.5 *(Continued)*

Factor No.	Name of Factors	Label	Statements	Factor Loading
4.	Availability	S5	FMCG brands are available in your locality.	0.725
		S7	FMCG brands offer a wide variety of products.	0.996
		S20	FMCG brands fulfill the aspiration of all categories of consumers' requirement.	0.837

10.5.3 EVALUATING AND VALIDATING FACTOWRS THROUGH CONFIRMATORY FACTOR ANALYSIS

We have also employed CFA to substantiate the hypothesis that has been developed to support the findings of EFA. To explore various indices, we extensively relied on AMOS. The path model, generated through AMOS, helps to assess more meaningful relationships among observed variables by logically extending multiple regression models. Figure 10.1 illustrates the path diagram accordingly.

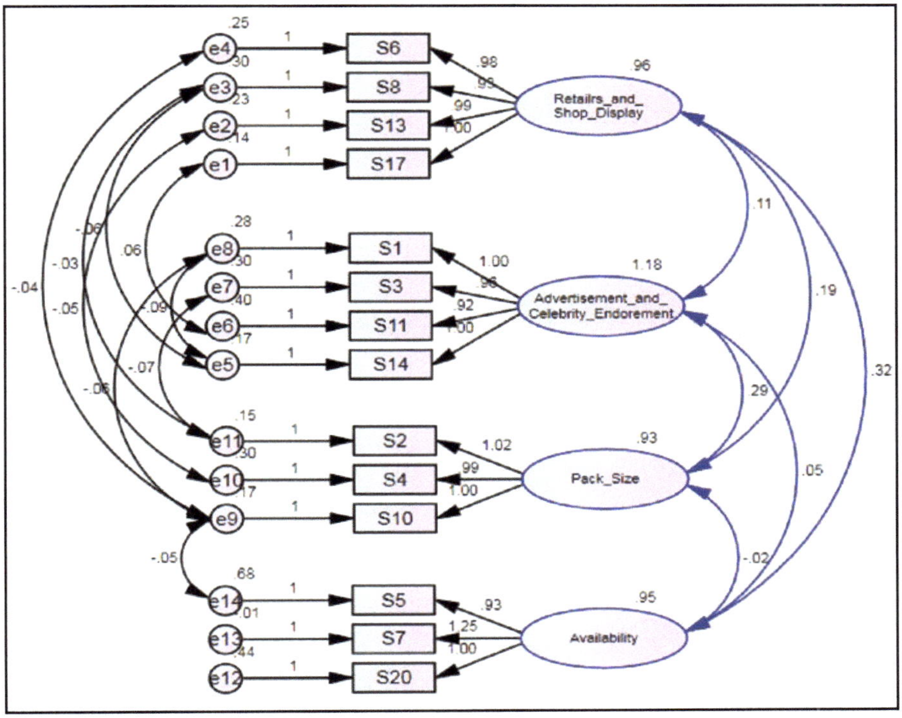

FIGURE 10.1 Path diagram.

AMOS sheds light on a considerable number of fitness indices to evaluate the goodness of model fit between the hypothesized model and sample data. Table 10.6 confirms the level of acceptable fit and fit indices for the proposed research model of four factors used in this present study.

TABLE 10.6 CFA Model Fit for Four Factors

Parameters/ Fitness Indices	Recommended Level of Good Model Fit Values	Values of Indices Obtained (Default Model)
Chi-square	–	79.443
DF	–	62
CMIN/DF	Should be near to 1	1.281
P	>0.05	0.067
GFI	>0.90	0.978
AGFI	>0.90	0.963
NFI	>0.90	0.987
TLI	>0.90	0.996
CFI	>0.95	0.997
RMR	<1	0.038
RMSEA	<0.05	0.024
PCLOSE	>0.5	1.000
PGFI	>0.5	0.577
ECVI	Default model < saturated and independence model.	0.332<0.421 & 12.172
FMIN	Should be closer to zero	0.159

In summary, our proposed model fits perfectly when considering fitness indices such as GFI, AGFI, CFI, TLI, NFI, FMIN, RMR, RMSEA, etc. Therefore, the model diligently fits in all aspects.

10.6 RESULT, DECISION, AND MANAGERIAL IMPLICATION

The results, decisions, and managerial implications of this study aim to draw inferences regarding how factors other than price and quality convincingly influence potential customers towards FMCG brands. This study will undoubtedly provide valuable insights and practical implications, considering the competitive nature of the FMCG market. Additionally, this study sheds light on the choice of media and its impact on the prospects of FMCG brands.

Based on the Thurstone Case V Scale (Table 10.3), it has been determined that audio-visual media remains the preferred choice for the target audience to gain in-depth knowledge about FMCG brands. With the revolution of JIO, the Internet is not far behind. People have become increasingly obsessed with the Internet for obtaining relevant information, and this holds true for the FMCG sector as well. Surprisingly, people now prefer the Internet even more than print media. The target audience seems to have a certain level of apathy towards broadcasting media.

During the process of factor analysis, we were able to extract four specific factors – advertisement and celebrity endorsement, pack size, availability, and retail and shop display. These factors, along with Price and Quality, have a significant influence in attracting the target audience towards FMCG brands. Marketers must have a deep understanding of these four factors in order to develop a well-thought-out marketing strategy that can prove its value in this highly complex and volatile market, characterized by constantly changing income patterns and buying behaviors (Patil et al., 2019).

The findings of this research highlight the importance for marketers to have a comprehensive understanding of these factors, as well as the choice of media, in order to effectively persuade prospects towards FMCG Brands. The insights provided in this study present a new opportunity for marketers to capitalize on and achieve significant success in the business by establishing themselves as trustworthy FMCG brands to both existing and new customers.

10.7 FUTURE RESEARCH DIRECTIONS

The present study will undoubtedly yield new insights, potential implications, and fruitful recommendations for academicians, decision makers, and practitioners. In addition to price and quality, we have managed to explore certain factors that significantly influence prospects towards FMCG brands. Furthermore, we have extracted the ranking of media channels for effectively reaching the target audience and informing, persuading, and reminding them about the brands. This study focuses specifically on the cosmetics and toiletries segment, with fairness cream, shampoo, soap, and deodorant being the chosen product categories for research. Future researchers can expand on this by highlighting other segments to provide more extensive and effective information. It is important to note that this narrow spectrum does not fully generalize the FMCG sector.

Moreover, the success of an organization heavily relies on the involvement of stakeholders and their knowledge, skills, and experience. However, this study does not incorporate the role of stakeholders and their influence in exceling the business. A detailed study would be necessary to explore the true nature of stakeholders and their impact on brand acceptance in a competitive environment. Therefore, this study provides a foundation for future researchers to conduct empirical studies on the role of stakeholders.

Like any other research work, this study has also encountered certain limitations. The study is confined to a specific time period, making its validity applicable only within this timeframe. The preferences and tastes of customers are rapidly evolving due to increasing income and the influence of Western culture (Tauseef, 2011), which suggests that predictions based on this study may not yield fruitful results. Additionally, the geographical scope of this study is limited to West Bengal only. As this study was conducted solely in Bengal, a promising state in India, scholars may question and be uncertain about the outcomes, as the results may significantly differ if the study were conducted over a longer period and in a larger geographical area. We employed simple random sampling to select the respondents, with a total of 440 participants. It would have been preferable to have a larger sample size, as the sample size has a definite impact on the study (Chander, 2017). The outcome of this study relies entirely on Primary Data, which was obtained through a structured questionnaire. In this study, we focused solely on four FMCG product categories (Fairness Cream, Shampoo, Deodorant, and Soap) within the cosmetics and toiletries segment. The results would be more rational and logical if we had managed to include other segments of FMCG products.

KEYWORDS

- **Ad media**
- **factor analysis**
- **FMCG**
- **key factors**
- **target audience**
- **thurstone scale**

REFERENCES

Ali, M., Thumiki, V., & Khan, N., (2012). Factors influencing purchase of FMCG by rural consumers in South India: An empirical study. *International Journal of Business Research and Development, 1*(1), 48–57. doi: 10.24102/ijbrd.v1i1.138.

Barclay, J. E., & Weaver, H. B., (1962). Comparative reliabilities and the ease of construction of Thurstone and Likert attitude scales. *J. Soc. Psychol., 58*, 109–120.

Berkowitz, D., Allaway, A., & D'Souza, G., (2001). The impact of differential lag effects on the allocation of advertising budgets across media. *Journal of Advertising Research, 41*(2), 27. https://link.gale.com/apps/doc/A76639398/AONE?u=anon~451b9b9f&sid=googleScholar&xid=ddab6de4 (accessed on 09 March 2024).

Bezjian-Avery, A., Calder, B., & Iacobucci, D., (1998). New media interactive advertising vs. traditional advertising. *Journal of Advertising Research, 38*(4), 23–32.

Bowling, A., (2002). *Research Methods in Health: Investigating Health and Health Services.* (2nd edn.). Buckingham: Open University Press.

Bulle, F., (2020). *The Effect of Brand Name on Consumer Buying Behavior: Empirical Study on Somali Students in Istanbul.* Available at SSRN: https://ssrn.com/abstract=3660519 or http://dx.doi.org/10.2139/ssrn.3660519 (accessed on 09 March 2024).

Chander, N. G., (2017). Sample size estimation. *The Journal of Indian Prosthodontic Society, 17*(3), 217, doi: 10.4103/jips.jips_169_17.

Chatterjee, K., Adhikary, K., Sen, S., & Kar, S., (2018). Identification and analysis of factors affecting consumer behavior in fast moving consumer goods sector. *Business Perspectives, 17*(1), 2–17.

Choi, S. M., & Rifon, N. J., (2012). It is a match: The impact of congruence between celebrity image and consumer ideal self on endorsement effectiveness, *Psychology & Marketing, 29*, 639–650.

Companies Listed on NSE (2019). *International Journal of Trend in Scientific Research and Development (IJTSRD)* (pp. 82–87). ISSN: 2456-6470, Special Issue | Fostering Innovation, Integration and Inclusion Through Interdisciplinary Practices in Management. URL: https://www.ijtsrd.com/papers/ijtsrd23070.pdf (accessed on 09 March 2024).

Cormack, D., (2000). The Research Process in Nursing. Oxford; Malden, MA: Blackwell Science.

DeCoster, J., (1998). *Overview of Factor Analysis.* Retrieved from: http://www.stat-help.com/notes.html (accessed on 09 March 2024).

Deliya, M., (2012). Consumer behavior towards the new packaging of FMCG products. *Journal of Research in Commerce and Management, 1*(11), 119–211.

Doyle, P., & Saunders, J., (1990). Multi-product advertising budgeting. *Institute of Operations Research and Management Science, 9*(2), 97–113.

Ducoffe, R. H., (1995). How consumers assess the value of advertising. *Journal of Current Issues & Research in Advertising, 17*(1), 1–18.

Edwards, A. L., & Kenney, K. C., (1946). A comparison of the Thurstone and Likert techniques of attitude scale construction. *Journal of Applied Psychology, 30*(1), 72–83. https://doi.org/10.1037/h0062418.

Farooq, W., & Latif, A., (2011). How urban children process advertising message: Special reference to television advertising in Pakistan. *African Journal of Business Management, 5*(10), 3962–3974.

Fill, C., (2009). *Marketing Communications: Interactivity, Communities, and Content.* Harlow, England: Pearson Education Limited.

Gilaninia, S., Taleghani, M., & Karimi, H., (2013). Internet advertising and consumer behavior in the purchase of products. *Kuwait Chapter of the Arabian Journal of Business and Management Review*, *2*(11), 1–5. doi: 10.12816/0001254.

Jayaraj, T., (2011). *55th Annual Report of Registrar of Newspaper for India (RNI).* Retrieved from: http://www.mxmindia.com/2011/12/registered-publications-show-6-25-per-cent-growth-rni-report/ (accessed on 09 March 2024).

Joreskog, K. G., (1967). Some contributions to maximum likelihood factor analysis. *Psychometrika, 32*, 443–482. https://doi.org/10.1007/BF02289658.

Kavitha, T. C., (2010). A comparative study of growth, challenges, and opportunities in FMCG of rural market. *Interscience Management Review*, *3*(2), Article 3. doi: 10.47893/IMR.2010.1056.

Malhotra, N. K., Hall, J., Shaw, M., & Oppenheim, P., (2008). *Essentials of Marketing Research: An Applied Orientation* (2nd edn.). Pearson Education Australia, Frenchs Forest.

Mishra, V., & Vashiath, A., (2017). Is advertising medium an important constraint in consumer purchase intention: The theoretical foundation. *Journal of General Management Research, 4*(1), 38–45.

Narayana, K., Lakshmi, & Mathew, B., (2015). A study on brand awareness and preference of rural consumers towards FMCG goods in rural markets with reference to Chikkaballapur District. *International Journal of Business and Administration Research Review*, *1*(9), 51–59.

Patil, P., (2016). An overview of Indian FMCG sector. *Paripex-Indian Journal of Research*, *5*(2), 171–173.

Petersen, L., (2018). *Advantages and Disadvantages of Radio Advertising.* Retrieved from: http://smallbusiness.chron.com/advantages-amp-disadvantages-radio-advertising-40629.html (accessed on 09 March 2024).

Price, J. L., & Mueller, C. W., (1986). *Absenteeism and Turnover Among Hospital Employees.* JAI Press, Greenwich.

Rafaya, M. M., Mahalaxmi, R., & Padmaja, R., (2017). Consumer behavior towards FMCG products: A study with special reference to Rajapalayam. *Research Journal of Humanities and Social Sciences*, *8*(1), 69–80.

Raghunath, U., (2018). Fast moving consumer goods retail market, growth prospect, market overview towards FMCG market in Indian market. *International Journal of Emerging Technologies and Innovative Research*, *5*(9), 76–79.

Shubhangi, A. P., & Viraj, V. J. (2019). A Study on Equity Research of Selected FMCG Companies Listed on NSE. *International Journal of Trend in Scientific Research and Development (IJTSRD),* Conference Issue, 82–87, https://doi.org/10.31142/ijtsrd23070.

Somireddy, K., (2018). Impact of advertisements on consumers towards FMCG products: A study in Hyderabad City. *World Wide Journal of Multidisciplinary Research and Development*, *4*(5), 105–112.

Statista, (2016). *The Advertising Industry in India Statistics & Facts.* Retrieved From: https://www.statista.com/topics/2116/advertising-industry-in-india/ (accessed on 09 March 2024).

Sugrova, M., Sedík, P., Kubelakova, & Svetlíkova, V., (2017). Impact of the product quality on consumer satisfaction and corporate brand. *Economic Annals-XXI*, *165*(5, 6), 133–137. doi: 10.21003/ea.V165-27.

Tabachnick, B. G., & Fidell, L. S., (2007). *Using Multivariate Statistics* (5th edn.). Upper Saddle River, NJ: Pearson Allyn & Bacon.
Tauseef, A., (2011). The impulse buying behavior of consumers for the FMCG products in Jodhpur. *Australian Journal of Basic and Applied Sciences*, *5*(11), 1704–1710.
Thurstone, L. L., (1929). Theory of attitude measurement. *Psychol Bull.*, *36*, 222–241.
Vibhuti, Tyagi, & Pandey, (2014). A study on consumer buying behavior towards selected FMCG products. *International Journal of Scientific Research and Management (IJSRM)*, *2*(8), 1168–1182.

CHAPTER 11

Specialized Technical Education and Entrepreneurism: An Empirical Study on SELF Program Participants

INDRANIL BOSE

Vice President-Academics and Professor, School of Business and Economics, Adamas University, Kolkata, West Bengal, India

ABSTRACT

The skill for employment in leather fabrication (SELF) program, funded by the Ministry of Rural Development, Government of India (GOI) under the Swarnjayanti Gram Swarozgar Yojana (SGSY) scheme, Pradhan Mantri Mudra Scheme, and coordinated and technically supported by different private players led by Infrastructure Leasing & Financial Services Limited (IL&FS Limited), has had a significant impact on the lives of thousands of rural youths below the poverty line (BPL) across India. As of March 2020, nearly 6,000 students from 40 training centers across nine states are earning their livelihood through the skills learned in the program.

The purpose of this study is to explore the entrepreneurial spirit of the participants of the *SELF program* across India. A structured questionnaire was used to interview 300 participants of the program. Tools such as factor analysis and the Z test were employed to determine the significance of different factors promoting entrepreneurial spirit among the students. The primary survey and subsequent research revealed that family factors and the technical institutional environment do not play significant roles in inspiring entrepreneurial spirit among the surveyed participants. However, the factor of having a role model has been found to be a significant determinant.

Contemporary Digital Transformation and Organizational Effectiveness in Business 4.0.
Nilanjan Ray, Sudin Bag, & Tapas Kumar Chatterjee (Eds.)

The chapter also recommends some initiatives on the part of the promoters of the *SELF program* to incorporate new measures to instill the spirit of entrepreneurship among the students. Additionally, the chapter acknowledges certain research limitations and identifies future opportunities for further research in this direction.

11.1 INTRODUCTION

The leather sector holds a crucial position in the Indian economy. According to a report published by the Indian Leather Product Association (ILPA, 2011), the leather sector is one of the top 10 export earners for the country, contributing nearly 4% of India's total exports (ILPA, 2021). The Indian Leather Sector Network, in its annual report (ILSN, 2021), has estimated that approximately 2.5 million people are directly or indirectly employed in the sector across India. The same report projects a total employment creation of about 6.5 million by 2019 if the current growth rate of 12% per year continues. The Council for Leather Exports (CLE, 2019) estimates that India has the capacity to meet nearly 10% of the global leather requirement and expects India's leather exports to double by 2016. The significant growth prospects in this sector have led to an increased demand for a well-trained and skilled workforce capable of working at all levels within the leather industry.

The SELF program, funded by the Ministry of Rural Development, Government of India (GOI) under the SGSY scheme, *Pradhan Mantri Mudra Scheme*, and coordinated and technically supported by various private players led by *Infrastructure Leasing & Financial Services Limited* (IL&FS Limited), has had a significant impact on the lives of thousands of rural youths below the poverty line (BPL). As of March 2020, more than 8,500 rural youths from the BPL category have enrolled in the program at 40 different training locations across nine states of India, and 6,500 participants have successfully completed the skill development course (http://self.skillschools.com/).

The main objective of the *SELF program* was to create a skilled workforce for the rapidly expanding Indian leather industry. However, the program also aimed to encourage a new generation of leather business entrepreneurs, albeit in small numbers. This extended purpose of the program should not be completely ignored (Bag & Omrane, 2022). Although the number is small, it is worth noting that only around 50 participants who have successfully completed the program have started their ventures across India. While this may seem insignificant, its long-term implications should not be dismissed based solely on the number of participants.

11.2 IMPLEMENTATION STRATEGY OF THE SELF PROGRAM: AN OVERVIEW

The implementation process of the SELF program is based on fundamental role equations and priorities:

- The main implementing agency, IL&FS, prioritizes the identification of required skills for the leather industry, which is done through the involvement of industry representatives (training centers/partners at a later stage).
- The selection of the target population for training intervention is a joint function of the key sponsor, the Department of Rural Development and Agriculture (DRDA), GOI, and IL&FS.
- IL&FS coordinates the function of imparting training and placement support, along with partnering organizations from the sectors.

It has been reported that approximately 5,900 participants who have completed the course under the *SELF program* have been successfully absorbed by different partner organizations across India. While the current employment creation by the program falls short of the expected number, it has certainly made an impact in the sector. Many partner organizations such as Mallcom (Kolkata), Relaxo (NCR), Mochiko (Uttaranchal), TATA (Dewas), Lakhani Vardaan (Faridabad), Super Tannery (Kanpur), Homera (Kanpur), Model (Kanpur), Calico (Kanpur), Super House (Kanpur), MAS (Chennai), Bhartiya (Nellore), PY Industries (Dewas), KH Group (Ranipet), etc., have already recruited a significant number of successful trainees from the program (IL&FS, 2022).

11.3 BRIEF LITERATURE REVIEW

A brief review of the existing research on the factors influencing entrepreneurial spirit or tendencies is presented as follows:

➢ Bal (1998) delved into the issue of communities and culture in entrepreneurship development in India. The author studied the nature of entrepreneurship, patterns of investment, range of business activities, and participation of family members in Punjab state. It was found that family factors play a role in shaping entrepreneurial efforts in individuals.

- Krishna (2003) attempted to study the gap between theory and practical implications of entrepreneurship. According to him, a well-conceived training intervention program can enhance the spirit of entrepreneurship in people. In his review of literary contributions, he concluded that such programs improve the sustainability and growth potential of enterprises in the long run.
- Shetty (2014) made an attempt to study the entrepreneurial orientation in Indian executives. The objective was to measure the attitudinal characteristics of Indian entrepreneurs. A sample of 200 executives was drawn, and scales of attitude were identified using hierarchical cluster analysis. Cultural differences were found to be a factor in shaping entrepreneurship.
- Nair & Pandey (2006) tried to study the characteristics of entrepreneurs in India. According to Bag & Omrane (2021), factors such as the economic status of the family, age, technical education/training, and work experience in a similar field seem to favor entrepreneurship. Entrepreneurs, compared to the rest of the population, tend to exhibit a more innovative attitude. Bhandari (2006); and Sinha et al. (2022) have examined the intention for entrepreneurship among students in Indian colleges. They administered a 62-item questionnaire to first-year management students in Rajasthan. The study found that variables like luck and the desire to lead others had significant relationships with entrepreneurial spirit. Levenburg & Schwarz (2008) explored the level of interest in entrepreneurship among India's business school students. They investigated the impact of culture, education, and environment on inculcating entrepreneurial spirit in Indian students. The authors suggested that these factors play key roles but vary greatly across nations. They also noted that in recent years, efforts have shifted the national mindset regarding entrepreneurship among India's youth, who now show a higher level of interest. However, the researcher has not found any literature on the entrepreneurial tendencies of trainees in occupational training from the leather industry in India, despite the sufficient number of studies on the factorial influences on entrepreneurial spirit.

11.4 RATIONALE OF THE STUDY

The stated objective of the SELF program is the creation of a skilled workforce for the Indian leather industry. However, the role of such a program in developing a new breed of small entrepreneurs in the leather industry cannot be completely ignored. One of the major objectives of the SELF program has been the dissemination of certain skills to unskilled workers employed in the industry through trained participants in far-flung areas of the country as well (http://self.skillschools.com/). This may be possible in different ways, and one such possibility is informal training at leather units set up by the new breed of small entrepreneurs in distant locations not yet covered by the SELF program. It has also been observed that the rate of post-training careers in the path of entrepreneurship in the leather industry has remained significantly low. As already stated earlier, less than 1% of the successful participants have started their own small ventures until 2011, and most of the ventures have been started in manufacturing and fabricating businesses (http://self.skillschools.com/). However, the study has shown that the curriculum framework of the SELF program has been continuously modified since its inception, and the latest version includes elementary knowledge of entrepreneurship. The course curriculum is indicative in nature, and the trainees are expected to be given training under the designed framework of the course curriculum at the training centers and at the places of training partners. Furthermore, it has been noted from the brief literature review that no study has been conducted to date on the entrepreneurial spirit of technical trainees with a basic educational background in general, and specifically on the Indian leather industry. In this context, the research aims to identify the effectiveness of various factors in influencing the entrepreneurial spirit of the participants, as only a few of these trainees have actually become entrepreneurs. Additionally, the research appears to be highly effective as it has been conducted on a selected number of trainees from a specific skill development initiative like the SELF program. This program is highly structured, well-coordinated on a PPP model, and has the potential to make a significant difference in the lives of thousands of underprivileged youths with limited schooling backgrounds from across India.

11.5 RESEARCH OBJECTIVES

Based on the rationale provided above, the following research objectives have been established:

- To assess the level of entrepreneurship among participants of the SELF program throughout India.
- To identify the factors that influence entrepreneurial spirit among participants of the SELF program throughout India.
- To determine the role of the SELF program and its curriculum in fostering entrepreneurial spirit among participants of the SELF program.
- To propose recommendations for promoting entrepreneurial spirit among participants of similar programs.

11.6 HYPOTHESES

For the purpose of the research, several null hypotheses have been formulated and tested for significance to demonstrate the objectives in a scientific manner. These are as follows:

- There is no significant impact of the curriculum on shaping the entrepreneurial spirit among the participants of the SELF program across India.
- There is no significant impact of the environment of the specific training centers and training partners on shaping the entrepreneurial spirit among participants of the SELF program.
- There is no significant impact of role models on shaping the entrepreneurial spirit among the participants of the SELF program.
- There is no significant impact of family influence on shaping the entrepreneurial spirit among the participants of the SELF program.

The above null hypotheses have been tested and results have been obtained.

11.7 RESEARCH METHODOLOGY

11.7.1 SAMPLING AND DATA COLLECTION

To study the entrepreneurial spirit among the participants of the SELF program, a study was conducted in 2019–2020. The study included 300 participants from five major leather industry locations: Kolkata, Kanpur, Agra, Ranipet, and Dewas. The sample consisted of trainees who were scheduled to complete their training by the end of 2020, as well as those who had completed their training within the past two years. A self-administered and non-distinguished five-scale structured questionnaire, containing 15 statements, was distributed among 328 participants on a random basis with the assistance of training partners. The respondents were from Kolkata (69), Kanpur (126), Agra (61), Ranipet (52), and Dewas (20). A total of 302 responses were received at the end of the 7th month of questionnaire distribution. However, two incomplete questionnaires were rejected, resulting in the final inclusion of 300 questionnaires for analysis. Therefore, 300 respondents were selected for the purpose of interpretation and analysis.

11.7.2 DATA ANALYSIS AND INTERPRETATION

The collected data were tabulated in an Excel sheet and analyzed using techniques such as factor analysis, item-to-total correlation, and Z-test through SPSS 17. Factor analysis identified six factors that indicate the spirit of entrepreneurship among the participants of the SELF program. These factors are mentioned in Table 11.1.

TABLE 11.1 Factor Analysis (Constituent Factors with Factor Load)

Sl. No.	Variable No.	Factor No.	Total Factor Load	Factor Name
1.	VAR00001	1	3.607359	Career ambition
	VAR00013			
	VAR00015			
2.	VAR00007	2	3.230184	Risk and uncertainty
	VAR00012			
	VAR00014			
3.	VAR00010	3	2.153657	Family support
	VAR00011			
4.	VAR00002	4	2.061303	Drive to lead
	VAR00004			
	VAR00003			
5.	VAR00008	5	1.830315	Training center/training partner environment
	VAR00009			
6.	VAR00005	6	1.059736	Social factors
	VAR00006			

- **Factor 1:** Career Ambition: This study examined the factors influencing career ambition priorities of participants in the SELF program. Factor 1, consisting of ambition, long-term benefits of entrepreneurship, experience, and total factor load of 3.607359 with 9.48% variance, indicated that most participants were interested in obtaining a job in the industry while also being inclined towards entrepreneurship.

- **Factor 2:** Risk and Uncertainty: Factor 2, focusing on risk and uncertainty, included inherent risks in entrepreneurship, uncertainty factors, playing it safe, and had a total factor load of 3.230184 with 8.61% variance. It was observed that participants considered risk and uncertainty as significant challenges in becoming an entrepreneur but acknowledged the highly rewarding returns.
- **Factor 3:** Family Support: Factor 3, related to family support, highlighted cases of failed entrepreneurial ventures and lack of support from family members. The total factor loading was 2.153657 with 6.73% variance. It was noted that only a few participants had received or were confident of receiving any support from their families in starting a new business.
- **Factor 4:** Drive to Lead: This factor consisted of an inclination towards entrepreneurship, putting ideas into practice, a desire to lead others, and the total factor load was 2.061303 with a calculated variance of 6.19. From the feedback analysis, it has been observed that most of the participants in the program intended to venture into entrepreneurship, wanted to put their creative ideas into practice, and lead others. However, due to a lack of support from family and a tendency towards risk aversion, they are deviated from becoming entrepreneurs.
- **Factor 5:** Training Center/Training Partner Environment: This factor consisted of the role of curriculum support and the impact of the training environment. The total factor load was 1.830315, accounting for 5.14% of the variance. Although the curriculum of the SELF program has been updated to include a small module on entrepreneurship, it has been found that most of the respondents have not been influenced by this modification.
- **Factor 6:** Social Factors: This factor consisted of entrepreneurs as role models and employers, with a total factor load of 1.059736 and accounting for 5.08% of the variance. The students consider successful entrepreneurs as role models, and it has been found that most of the program participants are influenced by the owners of the units where they are either undergoing training or have completed the training, both directly and indirectly.

Upon application of the Z-test formula, the first, second, and fourth null hypotheses have not been rejected at a significance level of 0.05, while the third null hypothesis has been rejected. Table 11.2

presents the results of the Z-test, including the table values and corresponding calculated values.

TABLE 11.2 Results of Z-Test as 0.05 Level of Significance

Null Hypothesis	Statement	Calculated Value	Results of Z-Test
1	No impact of curriculum in shaping entrepreneurial spirit.	1.225	Not rejected
2	No impact of training environment in shaping entrepreneurial spirit.	1.048	Not rejected
3	No impact of role models in shaping entrepreneurial spirit.	2.164	Rejected
4	No impact of family in shaping entrepreneurial spirit.	1.488	Not rejected

From the results of data analysis through z-test, it can be inferred that the course curriculum and the training environment do not play a significant role in shaping entrepreneurial spirit among the participants of the SELF program. As mentioned earlier, the aim of the course curriculum has been to mold the trainees into skilled employees for the leather industry in India. However, even though the SELF program recently incorporated an elementary module of entrepreneurship, its impact remained limited in promoting entrepreneurial spirit among the participants. Due to such factors, the majority of students (75%) prefer to take the safe route and join the queue of job seekers.

Regarding the significance of family influence in shaping entrepreneurial spirit, the analysis has shown that most participants have not chosen the path of entrepreneurship due to lack of family support, fear of risk, ambiguity, and fear of uncertainty. As for the impact of role models, an interesting trend is visible. It has been observed that most participants are inspired by the employers of their training units, but they are also fearful of any uncertainty.

11.8 CONCLUSION AND IMPLICATIONS

The study has many implications, particularly in relation to the discouraging situation regarding entrepreneurial spirit among the trainees of the SELF program. This program is symbolic, as there are numerous similar skill-building programs being conducted across the country, making a difference in the lives of thousands of literate and semi-literate youths from marginalized sections of society. Initially, most participants of the SELF program expressed their intention to start their careers as skilled employees in the leather industry, while some intended to start their own businesses at a later stage. However, the majority of trainees have expressed excitement about gaining inspiration from employers in the sector, as they have recognized the immense opportunities available for independent entrepreneurship.

In this context, the role of the SELF program becomes highly significant. Instead of solely relying on the natural inclination of trainees to become entrepreneurs or solely focusing on creating employability for thousands of fresh youths, the program promoters should begin instilling the spirit of entrepreneurship among the participants. To enhance the basic entrepreneurship module already included in the training curriculum, ILFS can seek assistance from national-level entrepreneurship promotion bodies such as the National Entrepreneurship Network (NEN) and the Entrepreneurship Development Council (EDC).

11.9 LIMITATION OF THE STUDY

As the study has been carried out in five locations and on a limited sample, the findings of the research may not be fully applicable in other locations as well. Moreover, all the feedback has been collected through single-source self-reported measures. Therefore, common method variance may not be denied to some extent. Responses of individual survey items may not be truly independent as there has been a chance influence of group membership. From these perspectives, it can be said that more research is needed to examine the generalization of this study on a larger sample size across more locations. A comparative study can also be done between different skill-building programs currently being offered for different categories of industries to get a more comprehensive picture of the issue.

KEYWORDS

- **data analysis**
- **empirical research**
- **entrepreneurial spirit**
- **national entrepreneurship network**
- **primary survey**
- **SELF program**
- **significant factor**
- **skill-building programs**

REFERENCES

Bag, S., & Omrane, A., (2021). The relationship between the personality traits of entrepreneurs and their decision-making process: The role of manufacturing SMEs' institutional environment in India. In: *Forum Scientiae Oeconomia* (Vol. 9, No. 3, pp. 103–122).

Bag, S., & Omrane, A., (2022). Corporate social responsibility and its overall effects on financial performance: Empirical evidence from Indian companies. *Journal of African Business, 23*(1), 264–280.

Bal, G., (1998). Communities and culture in entrepreneurship development in India. *Journal of Entrepreneurship* (Vol. 7, pp. 171–182). New Delhi.

Bhandari, N. C., (2006). Intention for entrepreneurship among students in India. *Journal of Entrepreneurship* (Vol. 15, pp. 169–179). New Delhi.

http://self.skillschools.com/ (accessed on 9 March 2023).

Krishna, K. V., (2003). Bridging the gap: Conceptual paradigms and training for entrepreneurship development. *Journal of Entrepreneurship* (Vol. 12, pp. 91–116). New Delhi.

Levenburg, N. M., & Schwarz, T., (2008). *Entrepreneurial Orientation Among the Youth of India, Firm Emergence and Growth.* UK: Emerald Group Publishing Limited.

Nair, K. R. G., & Pandey, A., (2006). Characteristics of entrepreneurs: An empirical analysis. *Journal of Entrepreneurship* (Vol. 15, pp. 47–61). New Delhi.

Report, (2009). *Council of Leather Exports.* New Delhi.

Report, (2010). *Indian Leather Sector Network.* New Delhi.

Report, (2011). *Indian Leather Product Association.* Kolkata.

Shetty, P., (2014). Attitude towards entrepreneurship in organizations. *Journal of Entrepreneurship* (Vol. 13, pp. 53–68). New Delhi.

Sinha, A., Bag, S., & Mandal, M. K., (2022). Intention of students towards tech-oriented teaching-learning method: A study of undergraduate and postgraduate level. *The Online Journal of Distance Education and e-Learning, 10*(2), 228–239.

CHAPTER 12

New Developments in the Agri-Food Industry: Manipulating Prospects for Industry 4.0

NADIA ADNAN[1] and HAFSA FAHAD[2]

[1]*Assistant Professor, Management and Marketing College of Business Administration, Prince Mohammad Bin Fahd University, Dhahran, Saudi Arabia*

[2]*UNSW Student, University of New South Wales, Australia*

ABSTRACT

In this chapter, we highlight opportunities for innovation in the agri-food sector. Traditionally, the agri-food sector has many inefficiencies that need to be addressed. The agri-food sector is crucial for human existence, and therefore, any efficiency gains can result in a drastic improvement in living standards across the world. According to estimates, more than 20% of the world's food produced is wasted across the supply chain. This lost food is then priced into the remaining food, resulting in higher food prices than they would be if we prevented those inefficiency losses. Additionally, the reduction in loss would aid the world in alleviating food poverty and food shortages.

Innovation can be brought about at the process level or product level. Innovation results in one or a combination of the following outcomes: increasing prototype speed, increasing output capacity, higher machine utilization, improved yield, increased customer satisfaction, process innovation, and increased customer satisfaction. In this chapter, we have explained six strategies specific to the agri-food sector that can be leveraged to achieve better production and distribution networks. The six strategies pertain to

Contemporary Digital Transformation and Organizational Effectiveness in Business 4.0.
Nilanjan Ray, Sudin Bag, & Tapas Kumar Chatterjee (Eds.)

advanced manufacturing technology, the internet of things (IoT), big data analytics, cybersecurity, horizontal and vertical integrations, and other miscellaneous strategies.

12.1 INTRODUCTION

Technology has expanded the frontier of possibilities in all sectors of the economy. It has not only led to improvements in the production process but has also created even more benefits in the distribution of output produced with the aid of technology. One key aspect of technology has been the improvement in the delivery process. Visibility across the value chain has increased and is further advancing, especially with the onset of the Industry 4.0 era. However, the full potential of technological advancements has yet to be fully exploited in some sectors. One key sector where technological prowess could be better utilized is the agri-food sector. The agri-food sector encompasses the value chain where food is processed, stored, transported, and delivered to end users. Unlike other sectors and economic outputs, the agri-food sector is characterized by perishability and sensitivity to a range of environmental variables. Even a small variation from optimal conditions can transform the end product into waste. The challenge in the food industry, however, is that controlling for too many factors at a granular level is cumbersome and almost impossible. Decades of technological advancements have helped the agri-food industry overcome its traditional inefficiencies. Specifically, technology has helped the agri-food industry in four different ways, all of which contribute to the efficient operation of the industry. The first way in which innovation and technology can help the agri-food industry is through advanced manufacturing machinery and products. Almost all agri-food products are created on a mass level, so there is great leverage for innovation at the machinery level. Even minor improvements in manufacturing machinery can create waves of impact. The second source of innovation in the agri-food sector is innovation in processes. Innovation at the process level can be achieved through superior information, expertise, or technology. The third source of innovation in the agri-food sector is data collection at various touchpoints. The agri-food sector needs to control many variables in the production environment and production itself. For example, temperature, proportions, moisture, and other ingredients are monitored to ensure desired outcomes. The internet of things (IoT) and related technologies assist in data collection at various stages of the production and distribution processes without human intervention. All of this data leads to better decision-making, timely response, and greater control over production processes. The fourth source of innovation is the distribution of the production to end users. As mentioned earlier, the majority of agri-food products are perishable and have a pre-defined shelf life. Therefore, it is crucial that the products are visible throughout the supply chain.

One major avenue of innovation in the agri-food sector can be the adaptation of technology from other sectors into the agri-food sector. This will increase the interoperability of the agri-food sectors with other industries, hence promoting increased efficiency. Technology from other sectors can easily be incorporated into the agri-food sector with minor tweaks. The application of technology in innovation can be more fruitful in the areas of manufacturing, distribution, and management.

In this chapter, we will discuss the applications of innovation and technology in the context of Industry 4.0 (also known as I4.0) for the agri-food sector. The rest of the chapter is divided into six strategies that can be implemented in the agri-food sector. Each strategy is defined with an example, implementation tactic, and a discussion on the success and failure of the strategy concerned. The strategies can be further dissected based on many attributes; however, we will not extend our discussion to that extent. For example, each strategy can be implemented for process-oriented innovation or product-based outcomes.

The chapter is then concluded with key outcomes summarized. We hope that this chapter will further our understanding about the possibilities of innovation in the agri-food sector in the context of I4.0 and beyond.

12.2 STRATEGY 1: ADVANCED MANUFACTURING SOLUTIONS

Innovation in advanced manufacturing solutions refers to the improvement of equipment and machinery used in production. As agri-food production occurs on a large scale, innovation in machinery and equipment can lead to significant cost and time savings, as well as better control over production. Additionally, making minor changes based on feedback from key customers can foster good relationships and ensure customer satisfaction. Manufacturing innovation can occur at both the product and packaging levels. Customers are increasingly concerned about the products they consume and are interested in understanding the entire value chain behind the end product. Consequently, packaging requirements have become more complex over time. Therefore, manufacturing innovation can result in improved customer satisfaction.

One example of a strategy is to develop a fermenter that reduces or eliminates the need for raw material storage time. Many food processing companies must follow specific procedures for optimal production. For instance, the Citrus firm mentioned in the sample case study needs to allow time for the fruits to ripen before processing. This requirement poses challenges for the firm, suppliers, and business vendors. The storage costs for the processing firm increase due to the need to store raw materials before processing. Moreover, the production time is lengthy and difficult to adjust as demand fluctuates. Introducing technology that reduces the storage time before processing could bring numerous benefits to the agri-food sector.

12.2.1 IMPLEMENTATION TACTIC OF STRATEGY

Implementation strategy can occur at various levels. Firstly, the strategy can be implemented where maximum impact can be achieved. For instance, if a key customer demands a change in product packaging, the company should invest in the necessary technology. Sometimes, customer requirements reach levels that cannot be met with existing technology. In such cases, new and innovative machinery is required. This machinery may be needed at any point along the supply chain. However, the agri-food sector can drive investments to locations that serve as bottlenecks, as mandated by clients, or that can result in maximum impact. Any decision regarding innovation made by management should be guided by the aforementioned priorities. The implementation of this strategy also needs to be well-planned, as innovation at the manufacturing level is costly and time-consuming. Therefore, managers should anticipate the useful life of any innovation before initiating the project.

There are several factors that can contribute to the success or failure of this strategy. We will discuss five main drivers of this strategy. If these factors are properly addressed, it is highly likely that innovation at the manufacturing machinery level will be successful. Since innovation at the manufacturing level often leads to product-based outcomes, it is crucial that such projects be executed with skill and caution.

Firstly, the impact of innovation at the manufacturing level shall be the primary criterion for evaluation. An innovation that generates a greater impact at the manufacturing level will have positive ripple effects throughout the entire supply chain. Therefore, innovation should be focused on high-impact areas where its utilization is maximized. In the context of manufacturing, innovation may also involve the customization of production machinery. This implies that the machinery should not be a monolithic structure, but rather composed of independent yet interoperable parts. By achieving granularity at the manufacturing level, the cost of customization can be reduced. Consequently, businesses will have satisfied clients and end customers. Moreover, manufacturing machinery, consisting of interoperable parts, can facilitate further innovations. This is because the cost of implementing an innovation will be low, and the potential downside of failure will be limited.

The second aspect of this strategy pertains to the likelihood of success for new projects. While it is ideal to assume the success of any project, innovation inherently carries risks. Therefore, it is the responsibility of management to assess the risk profile of each innovative proposition. As mentioned in the previous paragraph, if the manufacturing unit is composed of interoperable sub-units, the frequency of innovation can be increased while the risk of failure remains low. Hence, a key element in strategizing the innovation of manufacturing facilities is to divide the implementation into the smallest units possible.

The third key factor in innovation is the useful life of the innovation. Changing the manufacturing stack of an agri-food company is a time-intensive and capital-intensive project. Therefore, the usefulness of any project is one of the main considerations. It is not possible to change manufacturing equipment for every client's request. Therefore, agri-food businesses should strategize and prioritize changes based on pre-defined criteria. Two criteria that can guide businesses in this regard are industry-wide changes and key customer requirements. This means that the sources of initiating a change at the manufacturing machinery level should either be industry-induced or driven by principal clients. Any change that does not originate from industry trends or key accounts should be abandoned.

Dependence on parallel technology is another key theme that will contribute to the success or failure of any innovation project at the manufacturing level. A standalone change may be simple and less risky, but its impact will not be significant. On the contrary, a change that encompasses many units will be more impactful. This means that success will have better benefits if achieved, but the odds of failure will also be higher. There are many other minor factors that may contribute to the success and/or failure of any innovative project at the manufacturing equipment level, but the major drivers are those mentioned above.

12.3 STRATEGY 2: INTERNET OF THINGS (IOT)

The internet of things (IoT) is the second avenue through which the agri-food sector can reap tremendous benefits from Industry 4.0. IoT integrates technology and software with hardware products to create a unique combination of smart devices. Due to the numerous variables that need to be controlled in food processing firms, IoT is a revolutionary development that enables us to measure the production environment at a micro-level and make changes on the go according to pre-programmed scenarios.

In this strategy, IoT will be deployed across the supply chain to monitor the condition of food processing and final products, as well as maintain the desired level of environment. All of this will occur without human intervention. Primarily, IoT will enable food processing companies to perform three functions: data collection, task execution, and improved data provision. In addition to these functions, the potential of IoT is even more promising. It can embark on a new journey of mass customization and batching. Further explanation on these functions is provided in the explanation section.

An example of a strategy could be the implementation of IoT technology across the agri-food value chain. As mentioned in the chapter, many food processing companies require knowledge of the origin of their raw materials, from the farm to the packaging of the final product. For instance, meat-producing firms maintain comprehensive profiles of the animals that are slaughtered. Similarly, fruit processing companies need to keep track of data such as conditions, temperatures, and other variables. Furthermore, if there is a way to assess the conditions and initiate the necessary processes once the prerequisites are met, food processing can be automated. For example, in fresh fruit processing companies, the food is processed after a certain number of hours have passed or a specific temperature has been reached. With the assistance of IoT technology, data on the relevant variables can be collected, and the machine can be activated to begin the processing once the threshold conditions (such as hours or temperature) are met.

12.3.1 IMPLEMENTATION TACTIC OF STRATEGY

The implementation tactic of this strategy could be relatively simple. We need to perform three types of functions. Firstly, it is necessary to measure the variables of interest. For example, temperature, humidity, moisture, and air pressure need to be measured. The second aspect of IoT technology would be to utilize the gathered data for decision-making purposes and the implementation of pre-planned programs. For instance, a machine could be programmed to run once a certain threshold temperature is reached or a specific level of humidity is achieved. Similarly, the data measured and recorded with IoT technology can be utilized by management to make prompt and timely decisions. All of these functions can be integrated and consolidated into a single device. Moreover, IoT technology can operate in places where human intervention may not be possible. As a result, such strategies can yield significant benefits for food processing firms and the agri-food sector as a whole.

IoT technology has the potential to create a positive impact in the agri-food sector. The entire process of data gathering enabled by IoT technology produces valuable data that can lead to better insights. These insights can be directly utilized by individuals or can be used to train machines through modeling techniques. Machine learning (ML), for example, can be employed to leverage the micro-level data generated by IoT-enabled devices. This data can be used to simulate the production environment and automate as many processes as possible. By combining all of these steps, not only can the costs of production be reduced and production capacity increased, but the overall production process will also improve automatically. As the process continuously feeds its own data into the system with each iteration, better product outcomes can be achieved over time. However, the challenge lies in measuring the data that truly matters and disregarding the rest. When data measurement becomes easier, deciding what to measure and what not to measure becomes more complex. Therefore, it is essential to establish priorities for measuring data of interest.

Handling such large volumes of data is indeed a challenge. When production generates such a significant amount of data on a daily basis, it is crucial to ensure that the data is effectively utilized. Extracting valuable insights from the data necessitates proficient data wrangling skills. Consequently, big data analytics becomes a key factor for success. Another aspect to consider when dealing with IoT technology is the simultaneous measurement of numerous variables. While measuring more data may provide greater control over the production process, measuring an excessive number of variables can overshadow the main variables of interest.

Despite these challenges, there are also numerous opportunities presented by IoT technology. For instance, real-time and accurate data enables management to make informed decisions promptly. Enhanced decision-making, in turn, leads to improved profitability, reduced costs, and higher profit margins. Additionally, having access to more data allows for customized production tailored to the specific requirements of diverse customers. This, in turn, provides a competitive advantage and enhances sector efficiency and customer satisfaction. Customers are constantly seeking manufacturers and processors who can customize products according to their own brands. However, accommodating the rebranding requests of key customers poses a significant challenge for manufacturers as it requires adjustments to the production and packaging processes. For example, at times, key customers may request improvements in packaging or product cuts to enhance their appeal to customers.

Despite all these challenges, IoT technology can increase the visibility of the entire value chain and supply chain for suppliers, customers, and regulators. The food sector is one of the most regulated sectors of the economy. Therefore, regulations will also improve with more data availability. The cost of regulation will be reduced and so on. Customer awareness has been increasing for the past several decades. Customers increasingly want to know where their food has been grown and how it has reached them. They are also interested in knowing if climate-friendly practices have been followed. All of this information required by end customers can be collected, stored, provided, and utilized by IoT technology. More benefits of data will be discussed in the next strategies as well.

12.4 STRATEGY 3: BIG DATA ANALYTICS

Data is the new gold. The digitization of business processes has increased opportunities for businesses to obtain data on their operations that was previously unavailable. More data means better insights, which in turn lead to improved profitability. This strategy involves utilizing the data collected from various business processes across the value chain. The insights hidden within this data can be leveraged for numerous benefits, including reduced inventory costs, shorter production and lead times, lower storage costs, decreased wastage in production, and improved delivery processes throughout the value chain. However, to achieve these gains, we must uncover the insights hidden within the vast amount of data and translate them into practical and straightforward solutions. The effectiveness of these solutions directly impacts the savings that agri-food firms can realize. Overall, big data analytics presents a promising opportunity for agri-food businesses, and management should strive to harness these factors to their advantage. Companies often have a multitude of analytical options at their disposal, but they may become overwhelmed and default to a single approach (Rishi Bhandari, Marc Singer, and Hiek van der Scheer, June 2014). An example of a strategic application could be the planning of resource acquisition from suppliers. When there is elaborate data on the demand of a certain product, such as meat, processing companies can acquire raw materials in the right quantities at the right time without incurring excessive costs. This is only possible if the company's management has a strong demand model that incorporates seasonality and other factors. Similarly, processing companies can track customer inventories in their warehouses and storefronts. Armed with this key information, the processing firm can effectively produce and deliver the projected demand to the customer without any delays. This will enhance the business experience with processing companies and reduce inventory costs for both the processing firm and the customer, such as a supermarket. Likewise, data analytics can assist the company in managing its other resources, such as the delivery network and warehouses.

12.4.1 IMPLEMENTATION TACTIC OF STRATEGY

The gathered data will be divided into multiple sub-components. For instance, distinct data on relevant variables can be collected for the pre-production environment, warehousing, processing, delivery, and other areas. Subsequently, separate models can be employed for each of these areas to achieve continuous optimization of the processes. This will result in enhanced efficiency and responsiveness across all the key processes of the value chain. Management should strive to gather data for every key process of the value chain. Once the data is collected, it can be streamlined to uncover hidden insights. There are several elements that need to work together for an effective data analytics strategy. Firstly, priorities must be established for what needs to be measured and inferred from the data. These priorities should be based on the reality of the organization in question. For example, firms in the agri-food sector need to control processes such as inventory, raw materials, finished products, and customer delivery. These are the critical areas that require reliable data to refine the processes. Secondly, variables must be defined that can generate data aligned with the priorities set in the previous step. All variables of interest should be determined based on the mentioned priorities. For instance, inventory utilization can be one variable to measure the effectiveness of warehouse management. Similarly, the average time of inventory in the warehouse could be another variable to track the duration an inventory remains in the warehouse before being delivered to customers. Likewise, variables can be determined for other areas like raw materials, delivery, and other factors based on this data.

Once variables have been defined, the collection of data becomes necessary. It is imperative that the data is as accurate as possible and closely reflects real-time conditions. Given the fast-paced nature of the agri-food sector, data must be collected promptly. Once data has been collected on key areas, it must undergo a cleaning process. Irrelevant details should be filtered out to eliminate noise and ensure clarity. Following data cleaning, each area should be modeled separately. This could involve creating distinct models for warehouse management, production, raw materials, and delivery. Once the data has been modeled, the models should be refined through successive rounds of data analysis. The feedback loop should continue indefinitely to ensure that the model predictions closely align with real-world values.

The subsequent step involves deriving insights from the data and models to inform decision-making. These insights can contribute to reducing production time, lowering costs, or a combination of both. To achieve better results, there should be integration and interpolation of data across different areas. This means that the data and insights from one model should have defined relationships with the data and models from other domains. For instance, there should be a well-established relationship between raw materials and inventory, as raw materials are transformed into final products. Similarly, there is a clear relationship between inventory and demand. Over time, these relationships will become more representative of reality. By leveraging the analytics model, we can obtain prescriptive and descriptive insights from the data.

12.5 STRATEGY 4: CYBERSECURITY

Data is a crucial aspect of running a business, particularly in the agri-food sector. This is because consumers are highly conscious of their consumption choices and demand transparency in the food supply chain. However, achieving transparency requires a significant amount of data. This, in turn, poses a threat: the risk of data theft. Therefore, it is essential to take measures to protect the data.

The data security strategy primarily focuses on safeguarding the company's data for three main purposes. Firstly, if the data is vulnerable, it can be tampered with by malicious individuals. Considering that the production environment will be automated and integrated, it is crucial to implement security measures. Without these measures, the production process could be compromised. For example, let's say a snack requires 3 grams of chili per pack. If someone manages to hack the system and triples the proportion of chili, the entire production will become either useless or severely damage the brand. Similarly, if the login details of a key client are compromised, fake orders of significant quantities could be generated, leading to unwanted inventory.

The second motive behind data theft could be to obtain competitors' data. Agri-food business is a competitive industry, and the processes of each business are one of its key assets. Therefore, competitors might try to steal data from their rivals to gain knowledge about their processes or products. Additionally, end consumer data can be stolen for various purposes. Data theft has increased over the past decade, with many businesses such as banks and other companies facing such assaults on their data. Cyber-attacks have further increased after the COVID-19 pandemic. This is due to the increased instances of remote access to companies' assets, leading to more vulnerabilities and opportunities for hackers to steal data and carry out their malicious intentions. Data theft can be prevented through various strategies, including designing robust systems, training the workforce on system usage, and limiting privileged access to as few people as possible.

12.5.1 IMPLEMENTATION TACTIC OF STRATEGY

This strategy can be implemented through three main ways. Firstly, the digital system should be designed with limited vulnerabilities. The designs can be made robust in at least two ways. Firstly, data flow should be limited to the desired direction only. Secondly, users' access should be granted based on absolute needs. For example, privileged access should have special security protocols and should only be available to individuals with corresponding responsibilities. Thirdly, employees should receive training on the use of their systems. Training will enable employees to avoid falling victim to predatory tactics of hackers and other malicious individuals.

The cybersecurity of firms is primarily related to the IT department of the company. The IT department should make all necessary arrangements to secure the company's proprietary data and the data of its partners. If vulnerabilities are not addressed, untoward events can occur. Such events will tarnish the reputation of the concerned company and may trigger regulatory action against it. Therefore, it is crucial to safeguard all innovations from misuse through a robust mechanism.

12.6 STRATEGY 5: HORIZONTAL AND VERTICAL INTEGRATION

Horizontal and vertical integration are significant drivers of innovation. This type of innovation arises from the synergies that exist between different processes, divisions, or organizations. The strategy focuses on strengthening relationships between existing departments and business partners. Horizontal integration, for instance, involves integrating various processes within the same company. For example, the inventory management system can be linked to the sourcing system, which in turn can be linked to the accounting system, and so on. While these integrations provide fertile ground for innovation, the true innovations come from vertical integration. Vertical integration refers to integrating with the relevant systems of other business partners, both upstream and downstream. When combined, horizontal and vertical integrations not only leverage existing synergies but also foster greater creativity.

There are numerous strategies for integration. In the case of vertical integration, businesses and firms can integrate with upstream and downstream businesses across the supply chain. For example, a fruit processing firm may integrate its raw materials system with fruit farms. This integration would enable the fruit processing firm to obtain information on pest usage, water details, and other nutrient information. Similarly, fruit farms can integrate their raw material system with the processing firms to gain insight into inventory levels and make preparations for new sourcing. Likewise, the fruit processing firm can integrate its inventory system with the inventory system of its customers. This would provide visibility to the fruit processing firm's customers regarding inventory levels, expected demand, and more. Consequently, the fruit processing company can adjust its production to meet customer demand.

12.6.1 IMPLEMENTATION TACTIC OF STRATEGY

It must be noted that horizontal and vertical integrations should be approached strategically. Integrating with every software instance of the business partner is unwise for several reasons. Firstly, integrations can be costly, so more integrations mean more expenses. Secondly, excessive integrations expose business partners to unnecessary risks. Therefore, integrations should be limited to those that are necessary and avoid granting unwarranted permissions. The first challenge in implementation is to identify the priorities and goals that need to be achieved through integration. Horizontal integrations should also be kept to a minimum, although they are generally less risky than vertical integrations. Integrations provide efficient ways to access data that would otherwise be difficult or expensive to obtain. Integrations enable businesses and business divisions to seamlessly share data with each other.

Successful integrations reduce the costs of data gathering and overall business operations. For instance, in the case of meat processing companies, system integration allows each company to access inventory levels and other details of their counterparts. This eliminates the need for phone calls and manual communication. Instead, when animals arrive at the slaughterhouse with their tags and barcodes, the processing companies can simply tally those tags in their integrated system. The supplier system automatically provides all the necessary details of each animal to the processing company. The processing company can then modify the information according to their own practices and attach barcodes to the meat packets. Without integration, the supplier firm would have to bring paper documents to the processing company, resulting in increased costs and delayed business processes for both companies.

12.7 STRATEGY 6: OTHER ENABLING TECHNOLOGIES

We discussed five main strategies of innovation for the agri-food sector in the context of Industry 4.0. However, there are many more sources of innovation that go beyond these strategies. These innovations and their potential vary from case to case and depend on how they are implemented. Enabling technologies encompass various strategies, including custom software development, process-level innovation, and product-level innovation, among others. The strategies mentioned above in the sector only cover those related to infrastructure. However, innovations can also arise from improvements in production processes, warehousing, and distribution processes. Often, these process innovations are facilitated by some form of technology. However, innovation in processes may not necessarily be driven solely by technology. Other sources of innovation in process innovation are also possible. For example, process innovation can result from enhanced expertise, improved research, or a combination of these factors. Innovation can also arise from collaboration between two or more business partners. For instance, to expedite its delivery process, the Spanish meat processor (mentioned in the case study) could partner with third-party logistics companies to enhance its delivery capabilities. There are numerous examples of this strategy. Amazon, for example, has partnerships with top couriers like DHL, despite having its own delivery network. This is because Amazon recognizes that its delivery network may sometimes be overburdened or may not operate in certain areas. Therefore, Amazon has established backup options by partnering with other business entities. Such innovative strategies can be implemented by any business. Another potential innovation could involve sharing free capacities among businesses in the same industry. For instance, a processing firm that exceeds its production capacity could outsource some of its production to another company with available capacity.

12.7.1 *IMPLEMENTATION TACTIC OF STRATEGY*

Implementation of such strategies is a two-step process. In the first step, businesses will identify opportunities that can result in innovations. The second step will pertain to the implementation of the strategies and opportunities discovered. The proposed strategies may result in one or a combination of the following outcomes: increased prototype speed, increased output capacity, higher machine utilization, improved yield, increased customer satisfaction, process innovation, and increased customer satisfaction.

In the preceding section, we mentioned the possible targets of innovation. Increased production capacity, for example, can be achieved either through installing new machinery or through better use of the current solution. However, increasing capacity by installing more production machinery is not innovation. Innovation shall result in more production from existing machinery. This is possible only if the idle time is reduced, the speed of machines is increased, etc. Higher machine utilization can be attained through some innovation. For example, decreasing the downtime of a machine by automating its maintenance will increase the utilization of machines. There can be even more sources of innovation. Innovation, by its very nature, is not something that can be constrained or framed to some set of principles. Innovation means thinking outside conventions. The success of an innovation project is difficult to define and measure since it is multi-faceted and strongly depends on the expected output (Cooper & Kleinschmidt, 1987; Griffin & Page, 1996).

12.8 CONCLUSION

In this chapter, we have discussed innovation and the sources of innovation in the agri-food sector, specifically in the context of industry 4.0. We have primarily focused on six innovation strategies. Each strategy has been further elucidated with an example, its implementation tactic, and the factors contributing to its success or failure. The first strategy we discussed is innovation in manufacturing equipment. The second source of innovation is the IoT, which involves configuring software to hardware to automate certain processes. The third source of innovation is big data analytics, which serves as a breeding ground for innovation by providing insights into potential improvements. The fourth strategy of innovation is cyber security, which entails developing new ways to minimize data theft and protect a company's proprietary digital assets. Additionally, we explored the potential of horizontal and vertical integration as innovative mechanisms. Lastly, we discussed miscellaneous sources of innovation that can lead to process or product innovation.

The agri-food sector is of utmost importance in the economic landscape. This is because it covers the basic necessities required for survival. Therefore, it is crucial to ensure the timely processing and distribution of food and related products. However, there are concerns regarding inefficiencies in the agri-food sector. The most significant concern is the wastage of food throughout the supply chain. According to estimates, more than 10% of the food produced is wasted at various points in the value chain. This is a major problem that needs to be addressed through technological advancements. The two main sources of food wastage are inefficiencies in distribution systems and the lack of proper storage technologies. Therefore, significant innovations are required to reduce wastage across the supply chain. Further research is needed to contextualize existing technologies from different sectors and apply them to the agri-food sector.

KEYWORDS

- **agri-food sector**
- **big data analytics**
- **industry 4.0**
- **product-level innovation**
- **storage technologies**
- **third-party logistics companies**
- **vertical integration**

REFERENCES

Benitez, G. B., Ayala, N. F., & Frank, A. G., (2020). Industry 4.0 innovation ecosystems: An evolutionary perspective on value cocreation. *International Journal of Production Economics, 228,* 107735. https://doi.org/ 10.1016/j.ijpe.2020.107735.

Demartini, M., Pinna, C., Tonelli, F., Terzi, S., Sansone, C., & Testa, C., (2018). Food industry digitalization: From challenges and trends to opportunities and solutions. *IFAC-PapersOnLine, 51*(11), 1371–1378. https://doi.org/10.1016/j.ifacol.2018.08.337.

Garcia, M. M., Lazzarotti, V., Manzini, R., & Sánchez, G. M., (2014). Open innovation strategies in the food and drink industry: Determinants and impact on innovation performance. *International Journal of Technology Management, 66*(2, 3), 212. https://doi.org/10. 1504/ijtm.2014.064588.

Lawson, B., Krause, D., & Potter, A., (2015). Improving supplier new product development performance: The role of supplier development. *Journal of Product Innovation Management, 32*(5), 777–792. https://doi. org/10.1111/jpim.12231.

Lebersorger, S., & Schneider, F., (2014). Food loss rates at the food retail, influencing factors and reasons as a basis for waste prevention measures. *Waste Management, 34*, 1911–1919. https://doi.org/10.1016/j. wasman.2014.06.013.

Miranda, J., Ponce, P., Molina, A., & Wright, P., (2019). Sensing, smart, and sustainable technologies for agri-food 4.0. *Computers in Industry, 108*, 21–36. https://doi.org/10.1016/j. compind.2019.02.002.

Oltra-Mestre, M. J., Hargaden, V., Coughlan, P., & Segura-García, D. R. B., (2021). Innovation in the agri-food sector: Exploiting opportunities for industry 4.0. *Creat. Innov. Manag., 30*, 198–210. https://doi.org/10.1111/ caim.12418.

CHAPTER 13

Artificial Intelligence: The Road Map of Careers in the Financial Sectors

NADIA ADNAN,[1] AREEJ JAMIL,[2] and KUMAIL RAZA[2]

[1]*Assistant Professor and Research Scientist, Prince Mohammad Bin Fahad University, Management and Sciences, Saudi Arabia*

[2]*Researcher, Salim Habib University, Pakistan*

ABSTRACT

This chapter analyzes the effects of technological advancement, specifically artificial intelligence (AI), on various careers within the financial industry. It aims to gain insight into the evolving job market in light of these developments. The research begins with a comprehensive literature review, encompassing studies on the "efficient market hypothesis (EMH)," AI, and the impact it has had on different careers. To address the research question, interviews were conducted with experts in the relevant sector. The information gathered was then analyzed using a hypothetical framework to determine the validity of existing hypotheses or relational models. The study concludes that AI has indeed influenced careers in the financial sector. The flow of information has become significantly faster, requiring professionals in the financial industry to adapt at a faster pace. Some roles have become obsolete, with financial institutions now seeking individuals with expertise in Information Technology rather than solely economics or finance. As a result, there has been a shift in the demand for certain careers.

13.1 INTRODUCTION

The human species is considered the most intelligent creatures (Ehrlich, 2000) on Earth. Every form of animal or human is regarded as an intelligent

Contemporary Digital Transformation and Organizational Effectiveness in Business 4.0.
Nilanjan Ray, Sudin Bag, & Tapas Kumar Chatterjee (Eds.)

being to some extent (Brooks, 1991). Meanwhile, the Universe has evolved and developed in a way that the creatures inhabiting it are adapted to meet each other's existing and potential needs. Brooks argued that expertise, reasoning, problem-solving behavior, and language are the key factors in advancement. Furthermore, it is believed that humans are the most sophisticated intelligent beings in the Universe; our generations have become highly influential and commanding, especially in recent decades with the advent of computing technology and its various applications.

It is an accepted reality that human beings on Earth still have uncertainties about future predictions and analyses without concrete evidence (Moore, 2020). However, time has made it possible for computers, with the aid of technology, to analyze and examine available information and make future predictions based on certain assumptions (Moore, 2020). The situation where computing technology is capable of self-learning is known as ML. It is a subset of AI innovation frameworks (Mitchell, Michalski, & Carbonell, 2013). With the advancement of artificial intelligence (AI), computers can efficiently identify patterns better than human beings, making the predictions more reliable (Mitchell, Michalski, & Carbonell, 2013).

Furthermore, the advancement of technology has become an integral part of organizations. This has resulted in increased profitability for large companies, which are also known as technology-intensive industries. Reenen & Sadun (2010) claim that it is marginally less expensive to accomplish or access information, making it simpler to discuss endlessly with developing technology. For organizations in the financial industry to remain competitive, it is highly important to keep upgrading the technology by adopting it (Rahi & Ghani, 2018; Bag et al., 2021). Adding to this, Funnel (2021) also argues that with the evolution of the 5th generation, there is massive use of AI in every walk of life. ML is a significant resource when analyzing data and related matters. These mechanical machines are primarily viewed as promising in enterprises with much need for data analysis, like the financial sector (Ticknor, 2013).

In addition to that, the competitive advantage of technologies can be witnessed through their several operations; nowadays, companies prefer to set up machines to perform labor work instead of the human workforce (Autor, 2015). Such machines are expensive, but they have only one-time cost with minimum maintenance, contrasted with a workforce that needs to get paid each month. The aftereffect of implementing machines is that individuals can be replaced, like stock analysts (Li & Lu, 2020). Moreover, developing technology or machines has reduced prices compared to past years with better quality (Balsmeier & Woerter, 2019).

Nowadays, it is much simpler to operate or set up any business in the financial industry because boundaries have been eliminated, and now everything can be monitored virtually. This is something that organizations managing financial services (FinTech and Banks) have taken advantage of (Bofondi & Gobbi, 2017). As a result, they offer customers friendly arrangements at the lowest cost, leading the larger financial institutions or sectors to follow suit.

13.1.1 PROBLEM DISCUSSION

Data scientists worldwide and corporate professionals such as Jack Ma, Bill Gates, and Hawking have warned about the significant unemployment caused by innovation and technology in the industry (Brougham & Haar, 2013). However, this situation was anticipated with the advent of technology (Athey, 2019). Since the Industrial Revolution, also known as Revolution 5.0, machines have been replacing human jobs, leading to adverse effects on individuals' lives in the form of unemployment and rising inflation. The past two decades have been dedicated to the Technological Era of Shift, where the development of AI models has been introduced. These technological advancements, particularly AI, are gaining more popularity. They can perform a large portion of human activities at significantly lower costs. However, office occupations in the financial sector have been particularly affected. Consequently, highly educated individuals may become redundant in the future (Dhanabalan & Sathish, 2018). Recent studies have found a high level of anxiety regarding automation and AI trends, as people fear being replaced by machines (Athey, 2019). It is projected that as early as 2030, one-third to one-fourth of the existing professions can be taken over and replaced by machines, with a possibility of 50% of current positions being replaced within the next 10 years. Furthermore, recent research conducted by Bank of America and the University of Oxford suggests that 35% of jobs could be automated out of existence (Halal, Kolber, & Davis, 2017).

On the contrary, experts in AI and robotics, Varghese, state that AI cannot necessarily replace human labor (Horyat & Job, 2020). There is a debate among representatives about whether machines will take over all work, resulting in joblessness, which raises concerns about career development and job opportunities. Furthermore, Hagelbäck (an AI expert) guarantees that machines can deliver goods at a faster rate than humans and argues that this will lead to an increased demand for jobs in areas other than production, such as marketing or finance (Wimmerberg, 2017). Bessen (2016) argues that the advancement of AI will not be a threat to the job market; instead, it will lead to the evolution of existing jobs and the creation of new career paths for future generations. Additionally, the trend among organizations is shifting towards being publicly traded, which means there are more companies to analyze and more work to be done (Kumar & Sharma, 2017). Therefore, there is a need for machines to assist investment managers in making decisions (Bahrammirzaee, 2010).

Alluding to careers or jobs in the financial sector, individuals attempt to predict how the market will behave on a regular basis (Bekiros & Luecy, 2017). The financial market is highly complex and influenced by several different factors. To analyze the direction in which the market is trending, various types of analysis are commonly used, including fundamental and technical analysis. According to Markowitz (1952), the objective of a financial investor should be to maximize profits while considering the risk associated with the investment portfolio. AI, particularly neural networks, which are a part of AI, have enabled the industry to predict future price movements more efficiently than humans (Ticknor, 2013). Artificial neural networks (ANNs) have gained prominence in industries aligned with simulation modeling and are capable of making highly accurate predictions (Bontempi, Taieb, & Borgne, 2012). This is because machines are adaptable and capable of performing calculations and forecasts at a faster pace (Bahrammirzaee, 2010).

These new artificial machines are shaping the future of careers, and the job market is an interesting subject to study and examine, especially for individuals entering the market. Therefore, many positions that people are currently aspiring to may not turn out as expected due to ongoing technological advancements (Bessen, 2016). There is also a possibility that the ideal job may not exist in the future (Athey, 2019). This is a fascinating subject, especially for business leaders. If they implement machines to perform labor-intensive tasks, there is a possibility that people may refuse to purchase the products in protest (Francois, 2018). This indicates that these AI machines have a significant impact on the economy and employment opportunities in a cyclical manner (Athey, 2019). The constant change and advancement in technology are the underlying reasons for this issue. While there have been numerous studies on the intersection of Finance and AI, there is limited research on the impact of AI on careers in the financial sector. This study aims to fill that research gap. The purpose of this investigation is to develop an understanding of how newly developed technologies (AI) affect careers and future job prospects in the financial market.

- **Research Question:** How does the technological shift (artificial intelligence (AI)) impact careers and jobs in the Financial Sector?

13.2 THEORETICAL FRAMEWORK

13.2.1 FINANCIAL SECTOR AND MARKET

Financial markets are virtual or physical places where trading of securities, such as shares, bonds, and commodities, occurs. These financial instruments differ in terms of their intrinsic value, and the forces of demand and supply determine the price. Analysts play a crucial role in assessing price volatility and predicting future price movements. Therefore, they must evaluate several factors to examine the flow of financing. The efficient market hypothesis (EMH) assumes that financial information, whether symmetric or asymmetric, is accessible and available to every individual who wants to access it. This makes it nearly impossible to beat the market and engage in insider trading, as the price already reflects all available information. However, this hypothesis is further classified into three types: weak form, strong form, and semi-strong form (Lillo et al., 2012):

1. **Weak Form:** The prediction of future prices cannot be analyzed from the past. However, fundamental analysis can help in identifying abnormal returns. This form of EMH predicts that price fluctuations in the future are entirely independent variables from the investor's point of view (Roosi & Gunradi, 2018).
2. **Strong Form:** This form of EMH states that prices of securities reflect symmetric information simultaneously (Roosi & Gunradi, 2018). The restriction on insider trading is a characteristic of this form of EMH (Björklund & Uhlin, 2017).
3. **Semi-Strong Form:** This form of EMH is a hybrid approach towards price adjustments in the market. According to Fama (1970), individuals using any analysis tool cannot outperform the market's operational mechanism. This includes both fundamental and technical analysis.

13.2.2 ARTIFICIAL INTELLIGENCE (AI)

The ultimate objective of artificial intelligence (AI) is to address practical information processing problems and provide representation (Marr, 1976).

The program's algorithm is trained through data processing (Marr, 1976; Nilsson, 1980). In his research, Marr (1976) also noted that a key characteristic of AI is that once programmed, it cannot be undone. According to Brooks (1991), the true essence of AI is to emulate the role of human beings with machines. The concept and meaning of AI were initially derived from John McCarthy, who outlined it in his 1955 paper. McCarthy et al. (1955) defined AI as "every aspect of learning or any other feature of intelligence can in principle be so precisely described that a machine can be made to simulate it" (McCarthy et al., 1955).

McCarthy also attempted to use Machine Language to obtain results and assess accuracy and efficiency. In other words, the main purpose of the research was to test the hypothesis that AI can replicate the role of the human brain by effectively dealing with complexity (Brooks, 1991).

13.2.2.1 POPULAR EXAMPLES OF AI USAGE

1. **Natural Language Processing (NLP):** This is a component of understanding human language (Chowdhury, 2003). Specialists in the field of NLP gather information on how people interact with language on a daily basis, which further helps assess the performance of assigned tasks and analyze detailed interactions (Chowdhury, 2003). Additionally, the Spam Email feature is another example of NLP functioning (Chowdhury, 2003).
2. **Machine Vision:** This is another example of AI technology, which utilizes a camera for analysis. Unlike human eyesight, machine vision offers various functions similar to human functioning. Furthermore, as stated by Bulanon, Kataoka, Ota, & Hiroma (2004), Japan employs this technology in its agricultural industry to address the shortage of human labor for harvesting purposes.
3. **Robotics:** It is the most prominent example of AI with which we are familiar. Robots are capable of performing complex human tasks with greater accuracy and efficiency (Smith & Anderson, 2014). The invention of robots has also revolutionized numerous sectors.

13.2.3 MACHINE LEARNING (ML)

In the last three decades, machine learning (ML) has been regarded as the focal point for reality-based organizations worldwide (Jordan & Mitchell,

2015). ML is a branch of study in Information Technology, where data processing devices are developed to solve problems without the need for constant instructions. Jordan & Mitchell (2015) argue that this advancement in the field has opened up new opportunities for forecasting and predictions, which are particularly valuable in weather forecasting by experts.

Samuel A. is widely recognized as a pioneer in ML due to his significant contributions to the field. In 1959, Samuels developed an algorithm to design board game strategies for various presumed situations, marking a major breakthrough in ML. Furthermore, Halal et al. (2017) suggest that ML modeling can be achieved through learning and practice.

13.2.4 ARTIFICIAL NEURAL SYSTEM

An artificial neural network (ANN) is a branch of technological development designed to simulate the human brain and its processing capability. This foundation of AI deals with complex structure problem solving, which is nearly impossible according to statistical standards or human minds. Furthermore, a neural network comprises many interconnected processors known as neurons or, in some literature, nodes (Schmidhuber, 2014). These are the best available techniques for approximating or predicting values by running specific modeling (Schmidhuber, 2014). The core objective of this network is to reduce errors or rejection regions by applying several constraints and probability factoring.

13.2.4.1 FORECASTING WITH ARTIFICIAL NEURAL NETWORK (ANN)

The ANN model is popular due to its forecasting feature. This essential technology has had an impact on various sectors, including power generation, pharmaceuticals, and finance (Sharda, 1994). Additionally, Bontempi stated in 2013 that "forecasting, with the salient feature of consistency, has given Neural Networks an advantage over traditional forecasting techniques." Therefore, this technology is considered flexible, as the algorithm is trained on redundant historical data, enabling predictions of complex situations (Bontempi et al., 2013).

Furthermore, Neural Networks were initially introduced in the financial sector to predict corporate bankruptcy, business failure, stock prices, and foreign exchange rates under conditions of uncertainty.

13.2.5 ARTIFICIAL INTELLIGENCE (AI) IN FINANCIAL SECTOR

The application of AI can be observed across various levels of the financial industry. Programs and software utilized in the financial sector are primarily built upon the foundation of ANNs. When it comes to dealing with the Algorithm of AI for the Stock Market (which is widely regarded as the most uncertain part of the Universe due to its volatility), technical indicators are set aside as input (Ticknor, 2013). These technical indicators are essentially Moving Averages in terms of their data nature. Neural networks have the ability to predict future stock prices due to their planning, learning, summarizing, and self-coordinating attributes. Nowadays, stock exchanges in developed countries are employing Robots with Neural systems to function as analysts, handling various unforeseen scenarios (Vui et al., 2013).

13.2.5.1 PORTFOLIO MANAGEMENT

AI in the financial sector is used to manage investors' portfolios, which previously was handled manually. This technological shift has enabled the access of unbiased advice on investments (Bahrammirzaee, 2010). This system has also empowered managers or executives with a more productive approach towards achieving desired outcomes for individuals or organizations.

13.2.5.2 PREDICTION AND PLANNING

According to Ticknor (2013), predicting trends in the financial market is very complex due to its dependency on various factors such as governmental issues, market news, organization financials, and different indicators. Researchers in the last century focused on the correlation and accuracy of AI and Technological Advancements in the evolving Financial Sector (Nath & Kirley, 2007). Additionally, the ANN Hybrid model is an essential tool for predicting future analysis of uncertainty.

13.2.5.3 FINANCIAL TECHNOLOGY (FINTECH)

Financial Technology, commonly known as FinTech, has opened up numerous opportunities for researchers and entrepreneurs in the industry. FinTech primarily refers to the intelligent financial solutions that were previously addressed using traditional methods. FinTech-related products or resources do not necessarily incorporate AI, but AI plays a significant role in enhancing their viability and feasibility. According to Demertzis, Merler, & Wolff (2017), "FinTech has the potential to significantly transform monetary intermediation structures. It can disrupt existing intermediaries by introducing new business models empowered by sophisticated algorithms, resulting in lower costs and potentially improved customer experiences." Additionally, Bofondi & Gobbi (2017) state that FinTech offerings are designed to be user-friendly and cost-efficient, with algorithms at their core.

The impact of FinTech can be observed in various aspects such as settlement and payments, market support systems, and investment management (Bofondi & Gobbi, 2017). Furthermore, there is a noticeable shift in market trends, with consumers increasingly preferring online transactions and banking over traditional brick-and-mortar banks. FinTech and AI have also played a crucial role in supporting the infrastructure of the highly regulated financial sector (Demertzis et al., 2017). Financial institutions have embraced digitalization through payment innovations, influenced by technology-savvy companies like Google, Facebook, and Paypal (Bofondi & Gobbi, 2017).

13.2.6 DEVELOPING TECHNOLOGY AND IMPACT ON CAREERS/JOBS

The Green Revolution, leading towards the Industrial Revolution (4.0), has significantly transformed the dynamics of careers in traditional society. Companies are now more focused on maximizing production with limited resources, resulting in a reduction in the human workforce (Autor, 2015). Consequently, many individuals will find their jobs or careers left behind due to technological advancements. Middle-class and lower-middle-class jobs are becoming obsolete, raising concerns about the economic well-being of individuals, as highlighted by Autor (2015). He expresses particular concern about the displacement of skilled and semi-skilled labor, compared to historical trends. However, this trend has also impacted office jobs. Bessen (2016) similarly argues that the advancement of AI systems is posing a threat to middle-class occupations such as clerks, analysts, bookkeepers, and others. Furthermore, Beaudry et al. (2013) cite that the demand for jobs in various careers is declining, despite an increasing number of individuals graduating from the job market. This scenario raises significant questions about the future career opportunities for individuals.

According to Eister's (2015) research, in the 19th century, 51% of the workforce was employed in agricultural sectors, which has now reduced to only 2%. This demonstrates the impact of technological shifts on career and job opportunities within an economy. Companies are motivated by profit-making through downsizing, as highlighted by Bessen (2016). A prime example of how technological shifts have affected careers can be observed in grocery shops, where self-checkout systems utilizing AI technology have replaced workers previously assigned to this task. While this has positively impacted the company's profitability based on cost-benefit analysis, it has had a negative effect on individuals' careers in the job market (Brougham & Haar, 2013).

Upgraded technology and machine intelligence have rendered many existing jobs, and in a broader sense, careers, obsolete. However, on the other hand, these technological shifts have also given rise to numerous new jobs. In other words, these shifts have created industries within every sector, leading to increased demand for various career aspects. Autor (2015) provides a concrete example of how an Automated Teller Machine (ATM) has impacted the careers of bank cashiers in the financial sector. While it has affected their roles, it has also opened up employment opportunities for those who are skilled in operating and maintaining the machines. Bessen (2016) also agrees that the opportunity cost of the aforementioned scenario is balanced, meaning that while some positions are lost, new opportunities are created. If a task is fully automated, someone will lose their job; however, if the work is only partially mechanized, it leads to increased demand for labor (Bessen, 2016). Nevertheless, a contrary view cannot be ignored; Halal et al. (2017) states that there is a higher probability that this situation, albeit temporary, will result in more jobs in the market, thereby increasing career opportunities for newcomers. However, it may also lead to significant repercussions as technology advances to the point of being the most intelligent.

According to Bahrammirzaee (2010), "With the current trend in various sectors, the financial industry is also significantly affected as companies and businesses are now more inclined towards using machines and technology to perform tasks due to their efficiency and accuracy. Researchers commonly argue that investing in machines and technology, rather than relying on human workforce, is considered as asset-based financing rather than a liability burden." Several studies have examined the relationship between humans and machines in the financial market, particularly in the stock exchange (Kumar Sharma, 2017). However, the majority of industrialists and financial experts agree that machines are the ideal substitute for labor in order to allocate resources efficiently (Frey & Osborne, 2013).

13.2.7 JOB PERFORMANCE AND SATISFACTION

It is an essential milestone for any corporation to satisfy their workers with the company's objectives and operations (Judge & Weiss, 2017). Taylor's theory of motivation also claims that a workforce satisfied with their management or organization is likely to be more productive and beneficial for the corporation (Hidayat, 2020). Job satisfaction is highly correlated with job performance, according to Weiss et al. (2017). Thus, committed employees are often considered as the backbone of the company as they set the benchmark for others through their performance. However, incentivizing is an essential element in dealing with performance and satisfaction (Ćulibrk, Delić & Mitrović, 2018).

13.3 RESEARCH METHODOLOGY

13.3.1 DESIGN (RESEARCH)

The main objective of this research is to gain an understanding of how the advancement of technology, specifically "artificial intelligence (AI)," affects careers in the financial sector. The focus is on answering the question of how this impact occurs. To delve deeper into this matter, semi-structured interviews were conducted as the primary data collection method within the financial sector. This qualitative approach was chosen to gather opinions in a free-form style (Dani, 2021). Pereira, Santos, Sempiterno, Costa, Dias, & António (2021) argue that when it comes to understanding specific issues or conditions, the suitability of a qualitative approach is more valid compared to the quantitative research method. Graebner & Eisenhardt (2007) also highlight the significance of qualitative research, describing it as more applicable and relevant in the social context for examining various realities.

13.3.2 DATA COLLECTION

Essential theoretical concepts that were relevant to the study have been further investigated. In this context, interviews were conducted based on these theories to collect primary data. The gathered data from the interviews serves as the foundation for the analysis and drawing of conclusions in the study. The interviews, as a research tool, were conducted either in person or through online platforms. Online interviews offer time efficiency and help overcome geographical boundaries.

However, in the current scenario of COVID-19, digital platforms have become an integral part of work-life (Hyland & Gardner, 2021). Online interviews for data acquisition were conducted through Zoom and Google Meet. Furthermore, Salmons (2009) suggested that online or digitally conducted interviews should not be excessively long. The recommended duration for the entire process should be between 30 and 60 minutes, with a focus on avoiding redundancy.

The data collected through interviews, whether in person or through online platforms, has undergone an interpretation process by the team of researchers to examine the connection between perceived behavioral phenomena of the respondents. These responses have laid the foundation for empirical analysis and conclusions, with careful consideration of the sensitivity of the research region (Shirke, 2018).

13.3.3 SEMI-STRUCTURED INTERVIEW AND TRANSCRIPTION

Semi-Structured Interviews or Research are considered perfectly suitable for gaining a deeper understanding of the designated topic. In this research chapter, semi-structured interviews were conducted to examine the impact of developing technology, specifically AI, on careers in the financial sector (Garaue, 2015). Sandelowsk (2000) states that semi-structured interviews are conducted to ensure freedom of thought for every individual, without limiting their flow of thought on the matter under investigation. Therefore, the question under investigation is mainly open-ended, aiming to capture each respondent's individual view. Moreover, understanding respondents' philosophy is equally essential (Bryman & Bell, 2013); a few distinct individuals had reviewed the questions in this study prior to conducting the interviews. To enhance the trustworthiness or credibility of the topic under consideration, an Interview or Question Guide titled "AI & Careers in Financial Sector" is enclosed (Appendix 1).

In simple terms, transcription involves writing down spoken words on paper (Bryman & Bell, 2013). Some semi-structured or digital interviews are intentionally recorded to ensure that no important points from the respondent are missed. In certain cases, recorded or available interviews are transcribed to validate or evaluate the response from a specific perspective (Bryman & Bell, 2013). Transcription, however, enhances the credibility of research findings.

13.3.4 SAMPLE

The group selected for the examination of the topic consisted of individuals from the Financial Sector, specifically chosen from various sub-classified careers. By focusing the study on a particular segment, it is possible to gain a deeper understanding of the matter being investigated (Bryman & Bell, 2013). The criteria for selecting respondents for the research were based on their experience in the sector and their current involvement in the Financial Industry (Bryman & Bell, 2013). All 10 respondents are employed in the financial sector of Pakistan, holding diverse positions within different companies. The perspective of the industry as a whole (Bryman & Bell, 2013). Respondents were requested to establish connections on the research topic through personal and professional networks (Table 13.1).

TABLE 13.1 Respondents Profile

Respondent Name	Gender	Age	Title/Designation	Experience in Financial Sector
Mr. Jamil Anwer	Male	58	Head of International Business	30 years
Mr. Ali Raza	Male	62	Internal Auditor	35 years
Ms. Gul-E-Erum	Female	49	Financial Analyst	21 years
Mr. Humaira Shakeel	Female	42	Portfolio Manager	12 years
Mr. Muhammad Haroon	Male	26	Stock Market Consultant	3 years
Ms. Kashifa Kashif	Female	37	Investment Specialist	15 years
Mr. Tariq Iqbal	Male	45	Financial Advisor	18 years
Mr. Ajwad Hasan	Male	32	Bank Manager	7 years
Mr. Abbas Wasaya	Male	49	Income Tax Expert	20 years
Mr. Ali Kanji	Male	67	Financial Expert	40 years

13.4 EMPIRICAL FINDINGS

13.4.1 ANALYSIS AND TECHNOLOGICAL DEVELOPMENT

Big financial institutions in the industry, such as banks, utilize modern techniques to perform fundamental, technical, and AI-based analysis. Out of 10 respondents, two respondents engage in making predictions through technical indicators. However, the rest of the respondents use a hybrid technique of analysis with the help of technology. Technical analysis combined with technology is used to enter the market, while fundamental analysis helps in trading securities. One respondent (Ali Kanji) stated that the essential element of investment is to understand the company's future performance in the financial sector. There was a consensus among respondents in their interviews that technology, specifically productive software, and constructed algorithms for the financial sector, has surpassed the role of analysts in the field. "Artificial Intelligence isn't a revolution; it is the evolution of Era" (Ajwad Hasan).

The majority of the respondents were delighted with the use of technology as it has increased their productivity and profitability. However, one of the respondents, Jamil Answer, claimed that there is a need for human interpretation along with technology as there are chances of technical errors with ML. Upgrading the technology has become the need of the time as consumers' behavior has also shifted towards the ease of conducting operations, as mentioned by Kashifa Kashif. All respondents were positively inclined towards the role of AI in delivering better product services, which has further made the heavy flow of information and operations possible. For example, technology development has enabled us to predict market fluctuations with 85% accuracy, as stated by Tariq Iqbal.

Respondents feel that corporations in the Financial Sector have no choice left but to follow Technological Advancements. However, one of the respondents stated that one of the problems associated with such an AI-based algorithm is hard to maintain or carry, which at a certain level becomes a hindrance, as mentioned by Ali Raza. On the other hand, another respondent enlightened the vitality of AI in opening the doors of the Financial Sector to the People of Other Disciplines. She further highlighted that the Advancement of Technology had taken the reach of financial services to every part of the world, as mentioned by Gul-E-Erum.

13.4.2 DEVELOPING TRENDS (CAREERS)

Companies in the financial industry are continuing to downsize, eliminating additional departments. In comparison, there is a growing demand for machines and technology as human employees are deemed too costly and prone to liability for the company. When discussing the Stock Market, which is the core of the Financial Industry, machines are utilized to analyze data by constructing and training algorithms. This trend has opened up new opportunities in the field of technology. Respondents have agreed that clerical and administrative roles, where non-specialists are primarily involved, will be replaced by the by-products of technology such as AI and machines. Some respondents expressed concerns about the role of humans in decision-making and managerial work within the financial industry.

Two-thirds of the respondents acknowledged that the skillset requirements and dynamics of a career in the Financial Industry have evolved over time. Nowadays, in order to have a promising career in finance, individuals need to possess a solid understanding of technology handling, particularly in AI and data management. This knowledge provides an advantage over those who are technically proficient in subject matters related to financial accounting and economics.

"A person with a relevant background of working in the financial sector for three years ago would not be employable without familiarizing themselves with the ongoing evolution in the industry," said Humaira Shakeel.

Banks in the financial sector, known as the backbone of the industry, are now prioritizing the employment of IT experts and individuals rather than experts in finance and economics, as machines have taken over those tasks with greater accuracy and efficiency. Respondents have agreed to hire a workforce capable of performing core work in finance, economics, and technology. This statement highlights the gap in the career or job market of the financial sector, which presents a significant opportunity for upcoming graduates or young professionals.

"Spending on human capital was past, spending on information and technology is present and future" (Ajwad Hassan).

The majority of respondents agree that the financial sector operates under perfect market competition, making it an ideal case study for examination. Within the sector, there is a belief that there will be disruptive changes in the way duties are performed in the financial industry. Large IT companies are expected to gain market share in the financial services industry by offering cost-efficient features and access to a large number of users. However, some view this competition as a significant threat to the existing industry.

13.5 ANALYSIS

13.5.1 ANALYSIS AND TECHNOLOGICAL DEVELOPMENT

According to Schumacher et al. (2009), individuals working in the financial market or industry are involved in predicting market behavior using various tools. Respondents have observed that they engage in making predictions through technical indicators, while others use a hybrid analysis technique with the aid of technology. Technical analysis combined with technology is used for market entry, while fundamental analysis aids in trading. According to Singh (2021), if the market mechanism, often referred to as Market Hypothesis, is weak, the probability of abnormal profit-making arises, but it violates the principle of equal opportunities for all. However, respondents' responses indicate that they are dealing with a semi-strong form of the market, utilizing technical analysis of securities along with technological assistance.

The implementation of AI in the financial sector has begun, which has transformed the working dynamics of the field (Bahrammirzaee, 2010; Bag & Islam, 2017). This statement also supports the empirical findings of the study, showing that the use of AI has made the job easier for employers and businesses. Simultaneously, it has enabled customers to have faster information flow and processing of their work. Moreover, Chesbrough (2010) states that adopting technological changes is the only way businesses can survive. Our empirical findings also support this idea put forth by Chesbrough.

Based on the empirical findings examined above, the efficient flow and processing of information have allowed financial services companies to save time and money (Bag et al., 2020). Bofondi et al. (2017); and Sinha et al. (2022) have also acknowledged that modern technological tools have made data and communication more accessible. As a result, we can observe the financial sector evolving and banks adopting technology to communicate and carry out various operations. Furthermore, businesses have always benefited from engaging with customers (Singh, 2021).

13.5.2 DEVELOPING TRENDS (CAREERS)

Empirical findings of the study indicate that even with the increasing digitalization and implementation of AI in the financial sector, there is still a need for finance professionals who directly engage in customer interactions and decision-making processes. Frey & Osberne (2017) emphasize the importance of human interaction, as there will come a point where human minds will no longer be involved. Furthermore, the requirements for a successful career in the finance sector have also changed (Bessen, 2016). A higher level of technological skills is now necessary, along with formal education, to thrive in this field (Roy et al., 2022). According to Farooqi & Negendra (2014), this shift in dynamics can be seen as a promising aspect of career development. If individuals feel that their competence aligns with the required skillset of the industry, it can lead to job satisfaction. However, Christen et al. (2006) considers it a significant factor affecting job performance. Empirical evidence suggests that the world is rapidly advancing due to technological advancements. Just two years ago, individuals who left their jobs without training would not remain employable in the same position (Hunjra et al., 2010).

According to the respondents, financial institutions or companies are reducing their employment and human resource (HR) share, and this trend will continue. With the advancement of technology and AI, several departments will become obsolete and be replaced by new ones (Frey & Osborn, 2013). Due to competition, companies and the entire financial industry are focused on providing efficient and cost-effective services to consumers, which can only be achieved through the use of machines (Bag & Omrane, 2022). This situation implies that highly educated or literate finance professionals may face challenges in finding desired jobs in the sector (Jouanjean, 2015). However, human features such as human interaction and value addition in an organization will continue to create opportunities for young and skilled professionals in finance.

13.6 CONCLUSION

The objective of this study was to examine the role of developing technology (artificial intelligence (AI)) on careers and jobs in the financial sector. To achieve a deeper understanding, a comparative analysis and theoretical framework were designed to examine the effects and impact of technology on the financial sector. The increasing usage of AI in businesses has posed a threat to the careers of experts and professionals in the field. Furthermore, this has significant implications at every level of work and is taken into account as a risk factor. The conclusion of the study suggests that the financial sector is growing and changing rapidly due to technological advancements. However, individuals in the financial industry are also becoming redundant.

- **Research Question:** How does the technological shift towards artificial intelligence (AI) impact careers and jobs in the financial sector?

The introduction of technology has significantly impacted careers and the job market in the financial sector. One critical factor is the faster flow of information, which has increased the pace of work. This has also placed a greater responsibility on finance professionals to stay updated with technology. However, it is important to handle AI with standard regulations to ensure data safety and security. On the positive side, the use of AI has made the financial sector more efficient and productive through various means. However, it has also had adverse effects on individuals' careers. Therefore, finance careers should now be planned to encompass a diversified range of qualities, skills, and specialization in the core discipline.

Research indicates that professionals in finance are increasingly inclined towards redundant behavior. Technology has shifted the demand from economics and investment experts to information and technology experts. On the other hand, this situation has also created numerous job opportunities and careers in the subsector of finance that align with AI.

13.7 STUDY IMPLICATION

This study has contributed to the development of understanding the impact of AI on the formal and informal financial sector or industry. It can be concluded that the ability of AI to learn and program itself has made it more promising than traditional forms of operations. Therefore, the implications of the study can be divided into three areas: social, managerial, and theoretical implications.

13.7.1 SOCIAL IMPLICATION

This contributes to the general public by highlighting the crucial aspects of careers and jobs in finance. On a broader scale, not only are careers associated with the financial industry affected by this innovation, but other disciplines are also facing its impact. For instance, advanced Programmed Technology in ANNs or Robots is replacing human positions and jobs in the field, resulting in increased unemployment.

Therefore, it is of utmost importance to understand and analyze the impact of AI on the community. Furthermore, policymakers and experts should take steps to comprehend the needs and dynamics of this disruptive innovation. Social implications may also encompass the impact of technology on the education and vocational training system in society.

13.7.2 MANAGERIAL IMPLICATION

The managerial implications of the study can be observed in the growing utilization of AI and ML algorithms, which have resulted in increased profit margins, reduced costs, and enhanced revenue generation for corporations. Furthermore, these technological advancements have improved efficiency and accuracy in processing and operations. On the other hand, this study also emphasizes the importance and necessity of human involvement in the creation and development of such problem-solving technologies.

13.7.3 THEORETICAL IMPLICATION

Over the past 50 years, AI has been extensively studied and examined from various perspectives. Research findings indicate that AI-based technology has a significant impact on careers and jobs across multiple sectors and industries. These findings also align with the existing literature on the subject.

The research findings confirm that even in the heavily regulated Financial Sector, AI technology remains at least at the Semi Strong Stage (Kumar & Sharan, 2017). Furthermore, the field of Finance or Financial Services requires continuous technological upgrades to maintain market share and ensure survival (Hopkins, 2014). Technological advancements enable increased productivity and provide better investment opportunities in the market (Zhang & Jin, 2017). Additionally, this research supports the findings regarding career-related concerns, as employees in the financial sector express uncertainty about being replaced by machines and technology (Bryman & Bell, 2013).

13.8 REFLECTION OF RESEARCH

To establish a comprehensive understanding of the research subject, the study employs the methodology of Qualitative Research. This approach allows the respondents to freely express their thoughts and feelings on the matter, without the constraints of transcription or interpretation. Therefore, a qualitative research tool was chosen as the impact of AI on the Financial Sector cannot be measured solely in numbers or numerically.

Essential information is gathered through Semi-Structured Interviews, conducted both in person and digitally. The formulation of interview questions (*see* Appendix 1) is based on theoretical concepts, ensuring their relevance to the subject matter. The empirical findings section presents transcribed excerpts from the interviews, providing a narrative for the research. However, it is important to note that the results may vary with a larger sample size.

The main objective of the study is to examine the impact of AI, specifically on the Financial Sector. Therefore, the findings should not be generalized to other sectors, as this could be misleading and vary across different industries. The focus of this examination is directed towards one specific industry. The results of the study have contributed to the cultural, theoretical, and managerial fields, with observations indicating a significant impact of AI on the financial sector.

KEYWORDS

- **artificial intelligence**
- **career**
- **financial market**
- **financial sector**
- **financial services**
- **jobs**
- **machine learning**
- **managerial implication**
- **social implication**
- **theoretical implication**

APPENDIX 1 – INTERVIEW QUESTION

Name of Respondent:
Title/Designation:

➢ **Question – Analysis and Technological Development:**

1. Which techniques of analysis do you prefer to estimate the expected return on equity?
2. How has technological advancement have made analysis or prediction easier, efficient, or accurate?

3. Do you think the technological Shift has brought positive energy to the company or worsened the situation?
4. How do you see Technological Development in the Field of Finance?

➢ **Questions – Developing Trends (Careers):**

1. How do you see careers within the financial sector due to Artificial Intelligence (AI) or Technological Shift?
2. How would you consider that your working environment is by the current advancement of innovation?
3. Is there any change in the required skill set for the career in finance, or is it following the same linear trend?
4. What is the potential of Jobs or People choosing a Career in Finance after 10 Years?

REFERENCES

Athey, S., (2019). The impact of machine learning on economics. In: *The Economics of Artificial Intelligence* (pp. 507–552). University of Chicago Press.

Bag, S., & Islam, S., (2017). Non-performing assets a biggest challenge in banking sector-a comparative study between India and Bangladesh banking sector. *ICTACT Journal on Management Studies, 3*(4), 620–624.

Bag, S., & Omrane, A., (2022). Corporate social responsibility and its overall effects on financial performance: Empirical evidence from Indian companies. *Journal of African Business, 23*(1), 264–280.

Bag, S., Ray, N., & Banerjee, B., (2021). Assessing the effects of experiential quality on behavioral intention of customers in banking services: The moderating role of experiential satisfaction. *FIIB Business Review*, 23197145211052817.

Bag, S., Ray, N., & Manna, A., (2020). Nexus between sustainable development and CSR—An empirical study on Indian nationalized banks. In: *Circular Economy and Engineering* (pp. 71–87). Springer, Cham.

Bahrammirzaee, A., (2010). A comparative survey of artificial intelligence applications in finance: Artificial neural networks, expert system, and hybrid intelligent systems. *Neural Computing and Applications, 19*(8), 1165–1195.

Balsmeier, B., & Woerter, M., (2019). Is this time different? How digitalization influences job creation and destruction. *Research Policy, 48*(8), 103765.

Bekiros, S., Jlassi, M., Lucey, B., Naoui, K., & Uddin, G. S., (2017). Herding behavior, market sentiment and volatility: Will the bubble resume? *The North American Journal of Economics and Finance, 42*, 107–131.

Bell, E., Bryman, A., & Harley, B., (2018). *Business Research Methods*. Oxford university press.

Bessen, J. E., (2016). *How Computer Automation Affects Occupations: Technology, Jobs, and Skills* (pp. 15–49). Boston Univ. school of law, law, and economics research paper.

Bofondi, M., & Gobbi, G., (2017). The big promise of FinTech. *European Economy*, (2), 107–119.

Bontempi, G., Taieb, S. B., & Le Borgne, Y. A., (2012). Machine learning strategies for time series forecasting. In: *European Business Intelligence Summer School* (pp. 62–77). Springer, Berlin, Heidelberg.

Brooks, R. A., (1991). Intelligence without representation. *Artificial Intelligence, 47*(1–3), 139–159.

Bulanon, D. M., Kataoka, T., Okamoto, H., & Hata, S. I., (2004). Development of a real–time machine vision system for the apple harvesting robot. In: *SICE 2004 Annual Conference* (Vol. 1, pp. 595–598). IEEE.

Cardon, P. W., Huang, Y., & Power, G., (2019). Leadership communication on internal digital platforms, emotional capital, and corporate performance: The case for leader-centric listening. *International Journal of Business Communication*, 2329488419828808.

Chesbrough, H., (2010). Business model innovation: Opportunities and barriers. *Long Range Planning, 43*(2, 3), 354–363.

Chowdhury, G. G., (2003). Natural language processing. *Annual Review of Information Science and Technology, 37*(1), 51–89.

Ćulibrk, J., Delić, M., Mitrović, S., & Ćulibrk, D., (2018). Job satisfaction, organizational commitment, and job involvement: The mediating role of job involvement. *Frontiers in Psychology, 9*, 132.

Dani, S., (2021). *Business Research Methods*.

Demertzis, M., Merler, S., & Wolff, G. B., (2018). Capital markets union and the fintech opportunity. *Journal of Financial Regulation, 4*(1), 157–165.

Dhanabalan, T., & Sathish, A., (2018). Transforming Indian industries through artificial intelligence and robotics in industry 4.0. *International Journal of Mechanical Engineering and Technology, 9*(10), 835–845.

Dietterich, T. G., & Michalski, R. S., (1983). A comparative review of selected methods for learning from examples. *Machine Learning*, 41–81.

Ehrlich, P. R., (2000). *Human Natures: Genes, Cultures, and the Human Prospect*. Island Press.

Eisenhardt, K. M., & Graebner, M. E., (2007). Theory building from cases: Opportunities and challenges. *Academy of Management Journal, 50*(1), 25–32.

Eister, A. W., (1965). Evalutions of selected jobs and occupations by university students in a developing Country: Pakistan. *Social Forces, 44*(1), 66–73.

Evans, M. E., Hodder, L., & Hopkins, P. E., (2014). The predictive ability of fair values for future financial performance of commercial banks and the relation of predictive ability to banks' share prices. *Contemporary Accounting Research, 31*(1), 13–44.

Farooqui, M. S., & Nagendra, A., (2014). The impact of person organization fit on job satisfaction and performance of the employees. *Procedia Economics and Finance, 11*, 122–129.

Fennell, D. A., (2021). Technology and the sustainable tourist in the new age of disruption. *Journal of Sustainable Tourism, 29*(5), 767–773.

François-Lavet, V., Henderson, P., Islam, R., Bellemare, M. G., & Pineau, J., (2018). *An Introduction to Deep Reinforcement Learning*. arXiv preprint arXiv:1811.12560.

Frey, C. B., & Osborne, M. A., (2017). The future of employment: How susceptible are jobs to computerization? *Technological Forecasting and Social Change, 114*, 254–280.

Graue, C., (2015). Qualitative data analysis. *International Journal of Sales, Retailing & Marketing, 4*(9), 5–14.

Gunn, S. R., (1998). Support vector machines for classification and regression. *ISIS Technical Report, 14*(1), 5–16.

Haar, J. M., & Brougham, D. M., (2013). An indigenous model of career satisfaction: Exploring the role of workplace cultural wellbeing. *Social Indicators Research, 110*(3), 873–890.

Hassan, M. R., Nath, B., & Kirley, M., (2007). A fusion model of HMM, ANN and GA for stock market forecasting. *Expert Systems with Applications, 33*(1), 171–180.

Hidayat, A. S., (2020). The impact of coaching and person-job fit on self-efficacy and performance. In: *Advances in Business, Management and Entrepreneurship* (pp. 706–708). CRC Press.

Horvat, T., & Job, J., (2020). The use of machine learning in sport outcome prediction: A review. *Wiley Interdisciplinary Reviews: Data Mining and Knowledge Discovery, 10*(5), e1380.

Hunjra, A. I., Chani, D., Irfan, M., Aslam, S., Azam, M., & Rehman, K. U., (2010). Factors effecting job satisfaction of employees in Pakistani banking sector. *African Journal of Business Management, 4*(10), 2157–2163.

Hyland-Wood, B., Gardner, J., Leask, J., & Ecker, U. K., (2021). Toward effective government communication strategies in the era of COVID-19. *Humanities and Social Sciences Communications, 8*(1), 1–11.

Jordan, M. I., & Mitchell, T. M., (2015). Machine learning: Trends, perspectives, and prospects. *Science, 349*(6245), 255–260.

Jouanjean, M. A., Massa, I., & Ve Velde, D. W., (2015). The role of development finance institutions in promoting jobs and structural transformation. *Development Finance Agenda (DEFA), 1*(1), 16–21.

Judge, T. A., Weiss, H. M., Kammeyer-Mueller, J. D., & Hulin, C. L., (2017). Job attitudes, job satisfaction, and job affect: A century of continuity and of change. *Journal of Applied Psychology, 102*(3), 356.

Kumar, A. K., & Sharma, S., (2017). Recent updates on different methods of pretreatment of lignocellulosic feedstocks: A review. *Bioresources and Bioprocessing, 4*(1), 1–19.

Li, C., Lin, A. P., & Lu, H., (2020). *Analyzing the Analysts: The Effects of Technical and Social Skills on Analyst Performance and Careers*. Available at SSRN 3656427.

Mitchell, R., Michalski, J., & Carbonell, T., (2013). *An Artificial Intelligence Approach*. Berlin: Springer.

Moore, G. E., (2020). Certainty. In: *Arguing About Knowledge* (pp. 462–465). Routledge.

Nelson, D. M., Pereira, A. C., & De Oliveira, R. A., (2017). Stock market's price movement prediction with LSTM neural networks. In: *2017 International Joint Conference on Neural Networks (IJCNN)* (pp. 1419–1426). IEEE.

Nerurkar, P., Shirke, A., Chandane, M., & Bhirud, S., (2018). Empirical analysis of data clustering algorithms. *Procedia Computer Science, 125*, 770–779.

Pereira, L., Santos, R., Sempiterno, M., Costa, R. L. D., Dias, Á., & António, N., (2021). Pereira problem solving: Business research methodology to explore open innovation. *Journal of Open Innovation: Technology, Market, and Complexity, 7*(1), 84.

Popa, D. C. S., Popa, D. N., Bogdan, V., & Simut, R., (2021). Composite financial performance index prediction–a neural networks approach. *Journal of Business Economics and Management, 22*(2), 277–296.

Rahi, S., & Ghani, M. A., (2018). The role of UTAUT, DOI, perceived technology security and game elements in internet banking adoption. *World Journal of Science, Technology and Sustainable Development.*

Ray, N., Mukherjee, T., & Bag, S., (2021). Chaos and complexity of understanding online shopping behavior from marketing perspective. In: *Chaos, Complexity and Leadership 2020* (pp. 131–145). Springer, Cham.

Rossi, M., & Gunardi, A., (2018). Efficient market hypothesis and stock market anomalies: Empirical evidence in four European countries. *Journal of Applied Business Research (JABR), 34*(1), 183–192.

Salmons, J., (2009). *Online Interviews in Real Time*. Sage.

Schmidhuber, J., (2015). Deep learning in neural networks: An overview. *Neural Networks, 61*, 85–117.

Schumaker, R. P., & Chen, H., (2009). Textual analysis of stock market prediction using breaking financial news: The AZFin text system. *ACM Transactions on Information Systems (TOIS), 27*(2), 1–19.

Singh, J. E., Babshetti, V., & Shivaprasad, H. N., (2021). Efficient market hypothesis to behavioral finance: A review of rationality to irrationality. *Materials Today: Proceedings.*

Sinha, A., Bag, S., & Mandal, M. K., (2022). Intention of students towards tech-oriented teaching-learning method: A study of undergraduate and postgraduate level. *The Online Journal of Distance Education and e-Learning, 10*(2), 228–239.

Smith, A., & Anderson, J., (2014). *AI, Robotics, and the Future of Jobs* (Vol. 6, p. 51). Pew Research Center.

Ticknor, J. L., (2013). A Bayesian regularized artificial neural network for stock market forecasting. *Expert Systems with Applications, 40*(14), 5501–5506.

Tiwari, R., Bastawade, D., Sharan, P., & Kumar, A., (2017). Performance analysis of reversible ALU in QCA. *Indian Journal of Science & Technology, 10*(29), 1–5.

Van, R. J., Bloom, N., Draca, M., Kretschmer, T., Sadun, R., Overman, H., & Schankerman, M., (2010). *The Economic Impact of ICT*. Final Report..

Vui, C. S., Soon, G. K., On, C. K., Alfred, R., & Anthony, P., (2013). A review of stock market prediction with artificial neural network (ANN). In: *2013 IEEE International Conference on Control System, Computing, and Engineering* (pp. 477–482). IEEE.

Wilson, R. L., & Sharda, R., (1994). Bankruptcy prediction using neural networks. *Decision Support Systems, 11*(5), 545–557.

Zheng, A., & Jin, J., (2017). *Using AI to Make Predictions on Stock Market.*

CHAPTER 14

Exponential Growth of Digital Ecosystems in a VUCA World and the Role of Information Management: A Global Market Perspective

SHARMILA KAYAL,[1] MAHUL BRAHMA,[2] and DHANONJOY KUMAR[3]

[1]*Associate Professor, Department of Journalism and Communication, School of Media and Communication, Adamas University, West Bengal, India*

[2]*Professor and Dean, School of Media and Communication (SoMC), Adamas University, Barasat, Kolkata, West Bengal, India*

[3]*Professor, Department of Management, Islamic University, Kushtia, Bangladesh*

ABSTRACT

The ongoing COVID-19 pandemic has proven to be a catalyst for the rapid growth of digital automated business on a large scale. It is expected that this trend will continue or even accelerate in the digital space beyond the pandemic. This study focuses on the impact of digital ecosystems on small and medium-sized enterprises (SMEs) in India, with the aim of exploring their approaches and prospects in coping with the challenges posed by the pandemic. The research analyzes the digital ecosystem in three dimensions – 'input,' 'production,' and 'output' and examines their role in the Indian market. Through intensive interviews with entrepreneurs, the researchers gathered insights into their plans for venturing into the digital ecosystem to enhance their businesses. The findings suggest that businesses need to

Contemporary Digital Transformation and Organizational Effectiveness in Business 4.0.
Nilanjan Ray, Sudin Bag, & Tapas Kumar Chatterjee (Eds.)

expand their operations and embrace various digital advertising techniques to make consumers comfortable with digital payment methods.

14.1 INTRODUCTION

Currently, the universal landscape is characterized by continuous growth and the utilization of digital data, playing a crucial role in shaping the economy of a country, regardless of its classification as developing, underdeveloped, or developed. It can be argued that in order to achieve exponential growth, the existing limitations of production and packaging structures will inevitably lead to an intangible marketplace with innovative solutions for factors such as input, production, and output in the SME sector. The business sector has witnessed remarkable progress in recent years, particularly in automation, which is unparalleled and captivating. The digital revolution is likely to result in increased employment opportunities, as it effectively combines the power of humans and machines. The virtual business ecosystem has also demonstrated significant impacts, capable of transforming consumer lifestyles. Consequently, the rise of the digital automated business industry is expected to lead to exponential growth. However, this innovative digital business ecosystem also presents challenges for individuals who seek to adapt and benefit from this new experience. The transition process may be arduous. Furthermore, it is anticipated that production methods will undergo drastic changes, owing to advancements in embedded technologies such as AI (artificial intelligence) and ML (machine learning (ML)), which may influence the preferred mode of conducting business. There is a strong inclination towards SMEs, which will contribute to economic equality and foster a more balanced digital society. According to data from the MSME Ministry as of May 16, 2021, India has approximately 6.3 crore MSMEs, accounting for nearly 29% of the country's GDP through both domestic and international trade (Tripathi, 2021).

14.2 MATERIALS AND METHODS

14.2.1 PROSPECT CHALLENGES OF DIGITALIZATION

In their study of how digital ecosystems are factors for open, distributed, loosely-coupled systems of processes, with a perfect blend of technology and manpower, it is ideal to build more agile organizations, additional flexible systems, and more adaptable societies (Soley, 2010).

The prospective challenges of digitalization, which can be seen in any business, have been identified by Busquets (2018). These challenges include new competitors, the Smart Home concept in current times, the online banking system (for transactions, b2b or b2c), and the demand for personalization. These challenges need to be addressed diligently in the digital mode.

By expanding the focus with an in-depth investigation consisting of flexible methods of development of trades in the automobile platform, the researcher has found that the expansions have substantial spillover effects on customer benefits in the competitive situation of automotive indemnification. These causes have been identified by Pütz, Murphy, Mullins, & O'Malley (2019).

Kopalle et al. (2020) have developed a framework for the progressive capabilities of legal businesses in aligning with online clients. This framework involves several steps, such as focusing on understanding clients to enhance virtual enticement, expanding selection and specifications through proper planning and examination, and engaging customers online by emphasizing desirable legal characteristics while limiting conflicting ideologies and intrinsic promotion. By adopting digital positioning, legal businesses can modify their approaches and effectively serve their clients.

Hoch & Brad (2020) discussed the implications of digitization in business models, which have significantly intensified in contemporary times. Their research highlights the importance of transforming conventional commercial representations to efficient online modes. Traditional businesses that have not embraced digital technologies need to reconsider their business models and strategies.

Arora & Rathi (2019) conducted a study to examine the issues and factors underlying the implementation of online/digital small and medium enterprises (SMEs). They divided the study into several elements, including both interior and peripheral features. The study found no significant correlation between the magnitude role of both the business organization and the administration inventiveness. Additionally, factors such as the dearth of backing from the hierarchies of the organization and the lack of competence in facilitating and accessing online modes were identified as obstacles in the business sector.

The on-demand and adaptable nature of cloud-based amenities is an evolving trait of technology. The adoption of cloud services comes with various issues that organizations and regulators need to address in order to project appropriate solutions for SMEs (Bhat, 2013).

Chatterjee et al. (2021) conducted a study that signifies the significant influence of apparent utility and diligent usage with an inclination for modification on business and trade in free SMEs online enterprises in India. Their research focuses on dual mediators that have substantial effects on associations with virtual trade free enterprise, assuming enticing outcomes.

Trung et al. (2019) conducted research on the implementation of digital business in the SMEs sector. They found that India, among other countries, effectively applied the concept of digital business to SMEs. The study highlighted the significant role played by the administrative part and its intervention in promoting and advancing digital business for SMEs.

In their empirical research conducted in Bellary, Karnataka, the authors – Biswas & Roy (2007) have presented findings that suggest a technology-driven solution can significantly enhance business growth and development in the Indian market, particularly for small and medium-sized enterprises (SMEs).

Bagale et al. (2021) observed and analyzed the applications of customer relationship management (CRM) with the concept of online transactions in SMEs. They found that this approach has become more feasible and transparent in online mode, leading to an increase in superiority. Furthermore, they stated that this possibility can benefit GDP when viewed from top to bottom.

14.3 SUMMARY OF RESULTS

The automation in digital business has its unique process to accelerate organizations' understanding of the true value of their data (big or meta data). This can help them attract potential customers and overcome the challenges associated with traditional analysis of meta/big data. An exclusive data discriminator can be effectively applied in the following niche areas:

- Distribute trailblazing data analytics resolutions;
- Descend adjoining real-time doable perceptions;
- Determine ground-breaking explanations to business difficulties;
- Get-up-and-go conversant decision-making thought process;
- Comprehend digital assets.

The present research aims to explore the factors underlying the conversion of conventional business preferences to digital modes for exponential growth among SME entrepreneurs. The consequences of digitization within the business sector have already witnessed significant growth in recent times. Given the ongoing COVID-19 pandemic, there is a strong need for the transformation of knowledge from conventional approaches to digital modes, with a practical perspective. The adoption of online or virtual skills, particularly in sectors that were previously not digitally equipped, can greatly benefit SMEs in redefining their business strategies.

A descriptive research design with a qualitative approach has been adopted for this study. An in-depth literature survey and intensive planned interviews have been taken into consideration. The aim is to establish a theoretical framework to assimilate the phenomena with the process. From a business perspective, four entrepreneurs of SMEs have been considered within an articulate context, with inputs backed by business model inputs in a diligent manner. An in-depth analysis has been performed to understand the real business scenario, and the findings have been evaluated by decoding precepts from their opinions.

14.3.1 DIGITAL INSIGHT MODEL

The present study has adopted a digital insights-driven business model to effectively understand the customer-centric economy. A significant amount of data can be sourced from various digital and other allied sources. This poses a challenge for organizations in the SME sector in India, as they need to address the concerns regarding digital business practices. Conventional systematic approaches are often considered the only viable means for them. With the ongoing pandemic, the integration of digital narratives with AI, ML, digital payment security, and IoT will greatly benefit business developers in countries like India. In this case, the deliverables in Table 14.1 can enhance their business in the digital ecosystem (Figure 14.1).

TABLE 14.1 Business Model with Digital Ecosystem

Business Incorporation and Renovation
Business Astuteness Unconventional Conception
Data Science and Advance Analytics
Information Architecture with the help of (digital information construction, leading information administration, information value, information ascendency)

Source: Conceptualized by the Author(s) (Schadler, 2015).

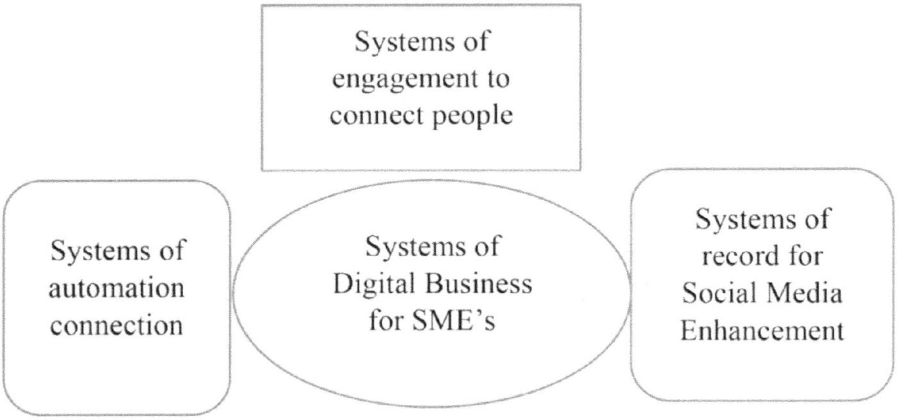

FIGURE 14.1 Business model with digital ecosystem.
Source: Conceptualized by the Author(s).

The concrete direction and aligned path are necessary for transitioning from clear understanding in order to obtain the virtual experience. What is needed most is joining, aggregation, storage, safeguarding, and handling with ease of dispensation of meta information, which proves to be a possible underlying factor for SME entrepreneurs. The aim of big data is to enable their alignment, a voyage of achievable commercial purposes for assessing the capabilities that should grow the competences to influence digital information prospects and also to generate information-rich resolutions of associated glitches, in addition to lifting proceedings as far as possible.

14.4 RESULTS AND ANALYSIS

All interviewees received a participant information sheet prior to the face-to-face interview. The interviews, which took place from August 2021 to October 2021, had a duration of 40–60 minutes and were recorded. The researcher, along with an interpreter, conducted interviews with a sample of four respondents. The participants were provided with a comprehensive explanation of the research objectives, and their input and experiences were sought to optimize the study. They were also given background information on digital business structures, including examples. The researcher supervised the interviews to ensure their relevance.

The in-depth interviews aimed to elicit participants' experiences and insights in three main areas: (i) the ease of accessing digital and physical modes of business for their own start-ups/ventures (e.g., their personal experiences with starting a business online and the subsequent transformation from physical to digital forms, and the reasons behind this paradigm shift); (ii) the challenges posed by the recent COVID-19 pandemic to their business or personal interests, particularly in the digital realm (e.g., difficulties faced in adapting to digital experiences); and (iii) the impact of digital advertising on their ventures (e.g., how competitors utilize digital ads for marketing and their own preference for digital automation in their ventures).

A qualitative analysis of the interview was conducted, involving coding, transcribing, labeling, and interpretation. After the interview, the information was categorized into various areas to provide a clear understanding of input, production, and output. The responses were then classified based on the relevant issues, such as digital advertising, promotion, payment, reach, and consumption. The role of digital ads can effectively enhance inclusive sustainability and customer engagement. The five main identified themes were further divided into sub-themes.

Following are the sub-themes:

1. **Attracting New Consumers:** The respondents confessed that they became aware of Internet usages for SMEs' own ventures through advertisements and promotions, which they learned about from referrals from their competitors. They are very interested in offering and learning about different payment options that are feasible for both the customers and themselves. Furthermore, they believe that with the new payment gateways and security measures, they will be able to attract a greater number of customers, as well as vendors and shareholders.

2. **Shifting Admittance in Digital Natives:** An upward progress in admittance from conventional forms of advertising and promotions to digital ones can be observed. The participants have acknowledged that customers are highly receptive to new digital initiatives and offers. They are pleased to receive services from the comfort of their homes, along with various offers and discounts, which is both welcoming and intriguing. The participants have also confirmed that the snowballing or referral technique, as well as positive word-of-mouth, have proven effective during this pandemic.

"I learned about digital business through a referral. Despite the lockdown, he was able to generate more money, which inspired me to follow suit. I decided to go for it and discovered the benefits of digital conversion for my own business. It worked very well for me.' (The interviewee's opinion)."

3. **Taking Up Additional Forms of Digital Mode of Business:** Marketing and preferment through business ads positively impacts the sustaining consumers related to SMEs.

   The entrepreneurs said;

   "It was by accident that I discovered the opportunity to invest in and explore the digital mode of communication, along with offering home delivery options to my existing customers. Through word of mouth, this concept gradually gained popularity, allowing me to generate a substantial income despite the lockdown. Additionally, there was no shortage of production."

4. **Taking Up Venture for the First Time Using the Digital Mode:** The respondents confessed about the pop-up advertising that usually comes up with their offers or ventures. The respondents did not like the enticements incited for the promotions, which caused a persistence of their attitudes.

   "I am unable to understand the advertisements; I don't know their meaning. I believe there are too many ads on the internet and social media... I dislike it because it manipulates my customers through alternative means of acquiring the product I am selling, while also giving competitors an idea about it and making it accessible to everyone. It constantly reminds everyone about my discounts/offers, and even if someone didn't have access to it, they might be able to forget about it. It's just impossible to have control over this."

➢ **Inputs:** With the digital automation of business, entrepreneurs gain several advantages that make it more feasible:

   - Raw material from all over India/globally in affordable prices;
   - Selection of suppliers (more suppliers can get in digital mode);
   - Able to choose from competitive prices;
   - Latest machinery/maintenance contacts.

They have opined that the production remains intact with greater ease of inputs, and the rate of production has significantly increased with a three-fold approach.

- **Outputs:** Areas of operation get increased because of digital automation of their business.
  - Advertisements (conventional);
  - Advertisements (digital);
  - Social media/social marketing;
  - Native and search advertising;
  - Mobile networks (devices-tools-connectivity-applications-analytics-personalization-security (two-step verification/mail verification/passwords and OTP) make it more transparent and digitally viable for both potential and existing customers;
  - Online order;
  - IoT censoring (refrigeration monitoring to push button alert system);
  - Collaborative technology;
  - BOPIS (buy online and pick up in store) concept;
  - Home delivery;
  - Subscription box for online orders;
  - Personalization through use of analytics (machine learning algorithm);
  - Artificial intelligence;
  - Payment gateway security;
  - Reverse auction for e-market place.

The results, based on the interpretation and observation from the intensive interview, suggest that marketing with advancements through digital ads facilitates easy access to inputs, production, and outputs, thereby increasing the number of customers among existing ones.

14.4.1 DISCUSSIONS

The small and medium-sized enterprises (SMEs) in India are increasingly adopting digital embedded technologies such as search ads, native ads, and social media marketing. These technologies help them interact with their customers and personalize their engagement. It is crucial for SMEs to embrace new-age technologies and practices to ensure smooth functioning.

This study aims to understand the usage and approaches of SMEs in the evolving landscape of technological preferences, in order to gain user-friendly insights. The study begins by identifying the factors associated with business and specific competences. It then explores how SMEs can leverage their proficiencies to generate innovative ideas for business growth. Additionally, it recognizes the importance of consumer perception and attachment to SMEs' entrepreneurial efforts. The study also identifies the factors that influence the progression of business competences and the understanding of consequent business results. Finally, it emphasizes the need to scrutinize digital literacy and the decision-making process of customers for the sustainability of SMEs in the Indian market.

14.4.1.1 ASSESSING THE DIGITAL THREATS

It is important to identify potential threats in digital automated business. Digital threats include hacks, frauds, and payment failures, which are vital to mitigate in order to integrate the digital mode for effective business. It is also important to understand the framework for digital analytics to gain insights into valued consumer behavior. External forces play a significant role in shaping the capabilities of SME entrepreneurs, which in turn influence decision-making for data and analytics-driven strategies in the digital ecosystem. Proper identification of value-based outcomes is crucial for generating insights.

14.4.1.2 WHAT'S NEXT?

An analysis of Strengths, Weaknesses, Opportunities, and Threats needs to be conducted in a comprehensive manner. Native and search advertising can greatly benefit the automation of the digital ecosystem for small and medium-sized enterprises (SMEs). The strategic utilization of search engine marketing is crucial for addressing user-centric issues.

14.4.1.3 DIGITAL DISRUPTION ON RISE!

The disruption in the digital mode is having a significant impact and presents a potential threat to automation in the digital ecosystem. Therefore, it is crucial to examine the potential challenges faced by digital automated SMEs, such as advertising blockers, rising advertising costs, lack of transparency for both sellers and buyers, ad blindness, and compliance with general data protection regulations.

14.5 CONCLUSION

The advent of digital technology has brought about a comprehensive growth in business and transformed the landscape of purchasing, particularly in the SMEs business arena. With the availability of up-to-date applications and the need for production inputs, operations have become more streamlined and accessible. Additionally, managing and communicating with existing or potential customers has become seamless, whether it is through mobile devices, smartphones, desktops, or computers.

14.5.1 LIMITATIONS AND FURTHER SCOPE

The research was conducted with selected SME entrepreneurs in India, focusing on intensive interviews with a restricted sample size. The aim of the study was to propose an integrated communication outline for adopting a theoretical framework, enabling them to navigate the digital landscape for their businesses. Additionally, this study provides an in-depth systematic analysis of the business approach in digital automation for exponential growth, which has not been addressed in previous studies.

Small and medium enterprises (SMEs) play a crucial role in India's economic progress and prosperity, constituting a significant percentage of the market. The concept of a free market has brought numerous benefits and opportunities for business revival and global connections. It is essential to recognize and support the significant role of SMEs. E-commerce and digital technologies, such as native advertising, search advertising, and social marketing strategies, can help mitigate challenges and enhance prospects for entrepreneurs, enabling them to compete with larger businesses in the online SME market.

There is a need for prerequisites that can facilitate modifications in light of advancements in modern technology. Embracing new innovations, customers play a crucial role in driving effective business and exponential growth, while also paving the way for new factors that can be explored through smartphone or computer-centric innovations for business development.

The catalyst for change can be attributed to advertising, which requires fast-paced communication and compliance with various digital strategies. This highlights the challenge of maintaining a balanced supply and demand ratio for sustainability, as well as the potential consequences of communication gaps. The emergence of digital automation technology necessitates efficient control of manpower and the avoidance of both physical and digital surplus, ultimately leading to transformative outcomes with inclusive support.

14.5.2 INNOVATION

The business sector diligently uses and innovates different means to connect with customers, vendors, and others using cloud-based technology. Innovation is crucial in recent times, especially during the COVID pandemic. Sustainability and brand positioning also require careful scrutiny with minimal investments.

14.5.3 INVOLVEMENT

Contribution includes incorporating sustainability practices in business by minimizing wastage of resources such as data, demand and investments, energy, and supply scale. This requires active involvement, proper attribution, and a focus on overall growth and prospects for larger benefits in the near future. Achieving this goal necessitates full commitment from all individuals associated with this endeavor.

14.5.4 MOTIVATION

A significant challenge faced by customers or buyers is the analysis and estimation of usage rates, as well as the resolution of accessibility issues. To address this, a unified business digital system has been implemented in the Indian market for SMEs. This system ensures continuous and specialized data management, optimizing speed and cost. Furthermore, it is essential to establish new user-friendly procedures that are accessible to all customers, regardless of their location, with an additional charge for the required components. The pricing should be set in a way that is affordable for all buyers or consumers.

KEYWORDS

- **digital disruption**
- **digital ecosystem**
- **exponential growth**
- **Indian market**
- **information management**
- **innovation**
- **small and medium enterprises**
- **volatility**
- **uncertainty**
- **complexity**
- **ambiguity**

REFERENCES

Arora, A. K., & Rathi, P., (2019). Determinants for the adoption of digitalization by manufacturing SMEs in India: An empirical study. *Prabandhan: Indian Journal of Management, 12*(8), 7–21.

Bagale, G. S., Vandadi, V. R., Singh, D., Sharma, D. K., Garlapati, D. V. K., Bommisetti, R. K., & Sengan, S., (2021). Small and medium-sized enterprises' contribution in digital technology. *Annals of Operations Research*, 1–24.

Barni, A., Brusaferri, A., Cavadini, F. A., Landolfi, G., Patil, S., Piga, D., & Vyatkin, V., (2019). Fostering the creation of a digital ecosystem by a distributed IEC-61499 based automation platform. In: *2019 IEEE 17th International Conference on Industrial Informatics (INDIN)* (Vol. 1, pp. 635–640). IEEE.

Bhat, J. M., (2013). Adoption of Cloud Computing by SMEs in India: A Study of the Institutional Factors. *Proceedings of the Nineteenth Americas Conference on Information Systems, Chicago, Illinois, August 15–17, 2013.* (Vol. 1, pp. 1–8). IEEE.

Biswas, S., & Roy, S., (2007). A shared ICT infrastructure for Indian SME clusters. In: *2007 International Conference on Information and Communication Technologies and Development* (pp. 1–11). IEEE.

Busquets, J., (2018). Multiasistencia: Transforming the digital ecosystem. *Journal of Information Technology Teaching Cases, 8*(2), 97–109.

Chatterjee, S., Chaudhuri, R., Vrontis, D., & Basile, G., (2021). Digital transformation and entrepreneurship process in SMEs of India: A moderating role of adoption of AI-CRM capability and strategic planning. *Journal of Strategy and Management.*

Digital Insights Services – Happiest Minds, (2017). Retrieved from: https://www.happiestminds.com/services/digital-insights/ (accessed on 09 March 2024).

Gupta, S., Leszkiewicz, A., Kumar, V., Bijmolt, T., & Potapov, D., (2020). Digital analytics: Modeling for insights and new methods. *Journal of Interactive Marketing, 51*, 26–43.

Hoch, N. B., & Brad, S., (2020). Managing business model innovation: An innovative approach towards designing a digital ecosystem and multi-sided platform. *Business Process Management Journal.*

Kopalle, P. K., Kumar, V., & Subramaniam, M., (2020). How legacy firms can embrace the digital ecosystem via digital customer orientation. *Journal of the Academy of Marketing Science, 48*(1), 114–131.

Pütz, F., Murphy, F., Mullins, M., & O'Malley, L., (2019). Connected automated vehicles and insurance: Analyzing future market-structure from a business ecosystem perspective. *Technology in Society, 59*, 101182.

Schadler, T., (2015). *Digital Insights Are the New Currency of Business.* ZDNet. Retrieved from: https://www.zdnet.com/article/emerging-technologies-to-power-your-systems-of-insight/ (accessed on 09 March 2024).

Solanki, V. K., Muthusamy, V., & Katiyar, S., (2016). Think home: A smart home as digital ecosystem. *Circuits and Systems, 7*(8), 1976.

Soley, R. M., (2010). Business ecology in the digital ecosystem. In: *4th IEEE International Conference on Digital Ecosystems and Technologies* (pp. 102–102). IEEE.

Tripathi, A., (2021). *MSMEs in India – Post Covid Scenario.* Times of India Blog. Retrieved from https://timesofindia.indiatimes.com/blogs/agyeya/msmes-in-india-post-covid-scenario/ (accessed on 09 March 2024).

Trung, N. X., Binh, D. T., Thuy, D. T., & Linh, D. T. T., (2019). Research on the application of e-commerce to small and medium enterprises (SMEs): The case of India. *Business and Economic Research, 9*(3), 102–113.

Weill, P., & Woerner, S. L., (2015). Thriving in an increasingly digital ecosystem. *MIT Sloan Management Review, 56*(4), 27.

Xiarewana, B., & Civelek, M. E., (2020). Effects of COVID-19 on China and the world economy: Birth pains of the post-digital ecosystem. *Journal of International Trade, Logistics and Law, 6*(1), 147–157.

CHAPTER 15

Medical Hotel: Innovative Strategy of Hospitality Industry for Survival and Growth

PRATIM CHATTERJEE,[1] JOSÉ G. VARGAS-HERNÁNDEZ,[2] DEBASISH BATABYAL,[1] and NAMRATA CHAKRABORTTY[3]

[1]*Assistant Professor, Amity University Kolkata, West Bengal, India*

[2]*Research Professor, Institute of Technological Postgraduate and Research Division, Instituto Tecnológico Mario Molina, Zapopan, Mexico*

[3]*Assistant Professor, Guru Nanak Institute of Technology, Kolkata, West Bengal, India*

ABSTRACT

With the growing importance of innovation for the survival and growth of the hospitality industry, as well as the emergence of healthcare hotels, the purpose of this study is to provide a literature review in this area. Additionally, this chapter aims to identify opportunities for future research.

The present study adopts the systematic literature review (SLR) method. A search was conducted on the Google Scholar search engine from January 2009 to December 2021, using keywords such as "Hotel Innovation," "Hospitality Innovation," "Organizational Innovation," and "Medical Hotel."

Existing literature indicates that limited research has been conducted in this area to date. Medical hotels have proven to be highly effective as both a hotel and a hospital, as well as an isolation center for medical tourists and patients. However, the potential of medical hotels as an innovative strategy in the hospitality industry remains largely unexplored.

Contemporary Digital Transformation and Organizational Effectiveness in Business 4.0.
Nilanjan Ray, Sudin Bag, & Tapas Kumar Chatterjee (Eds.)

This chapter reviews literature of a decade on Hospitality innovation and Healthcare hotel and is based on a very ignored yet potential strategy of hospitality industry. This chapter will be useful to hospitality professionals, researchers, and academicians.

15.1 INTRODUCTION

Innovation refers to any novel or new good, service, or idea. In order to provide exceptional customer value and surpass the expectations of their target customer segment, hotels must continuously adapt to changing customer desires and needs through innovation (Grissemann et al., 2013; Nagy, 2014). Similar to the manufacturing industry, new product development has become crucial for the survival and growth of the service industry. By consistently upgrading and innovating, firms can differentiate themselves from competitors in this modern era (Orfila-Sintes & Mattsson, 2009; Backman et al., 2017). Hotel category is an avenue for hotels to distinguish themselves from other properties (Becerra et al., 2013). Hotels need to conduct research and identify unique personal tastes and preferences in order to offer distinctive value-added and innovative products and services, alongside their existing offerings, to provide a memorable experience for customers (Bharwani & Mathews, 2016).

For hospitality and tourism organizations, innovation is a key strategy for creating new product and service developments and is considered the primary way to turn challenges into opportunities (Chen, 2011; Hillman & Kaliappen, 2015). In the third edition of the OECD Oslo Manual (2005, p. 46), four types of innovations are presented: (i) product innovations; (ii) process innovations; (iii) marketing innovations; and (iv) organizational innovations. The impact of external factors such as increasing tourist demand, competition, and changes in tourism and hospitality industry accelerates the need for innovation. An innovative hotel offers new products, services, processes, or ideas (Pavia & Floricic, 2017; Toivonen & Tuominen, 2009). Despite being considered one of the most competitive industries in the world, the hospitality industry experiences lower productivity compared to other industries, and the lack of innovation practices is responsible for this phenomenon (Martin-Rios & Ciobanu, 2019). Research on innovation and customer orientation in the hospitality industry is rapidly increasing, yet our understanding in both of these areas remains vague (Tajeddini & Trueman, 2012; Sandvik et al., 2014).

Medical hotels are conventional hotels that provide standard hotel services along with healthcare services to patients (Docrates, 2012; Medical Hotels, 2012). The transformation of the hospital sector to incorporate hospitality is crucial in order to enhance the customer experience (Zygourakis, 2014; Samal et al., 2017). Despite being the ideal destination for international inbound medical tourists, there is a scarcity of studies on medical hotels. The combination of medical treatment, accommodation, and food in a single establishment offers convenience to customers (Han, 2015; Han & Hwang, 2013). Unlike regular hospitals or medical clinics, medical hotels also provide accommodation for the family members of patients (Gan & Fredrick, 2011; Heung et al., 2011; Han, 2013). Medical hotels offer various facilities to patients and their families, including healthcare treatment, food, accommodation, safety, laundry, room service, and concierge services (GHN, 2012; Han & Hyun, 2014; Kazemha & Dehkordi, 2017). The medical functions of healthcare and the hospitality functions of hotels can easily be combined in a medical hotel, as both belong to the service industry (Jin & Zhang, 2016; Pavia & Floričić, 2017). Major hotel chains in collaboration with renowned specialist hospitals provide high-quality treatment, accommodation, and other essential services to ensure that patients experience more rest and less stress (Han & Hyun, 2014; Suess & Moody, 2017).

In any epidemic or pandemic, a medical hotel can be effectively utilized as a quarantine facility with proper medical and accommodation amenities, especially for patients who are unable to safely self-isolate at home (Ramírez-Cervantes et al., 2020; Hoefer et al., 2020; Haroon et al., 2020; Mackenzie et al., 2020). Hung et al. (2018) in their studies, highlighted the role of the hotel industry in responding to pandemics, epidemics, or any infectious diseases. A medical hotel provides an ideal environment for highly infected patients, offering a safe and secure setting with quality medical care provided by well-trained staff (Bannister et al., 2009). In special hotel-based ambulatory care units for critical patients, family members are also encouraged to stay with the patients whenever possible (Sive et al., 2012).

Therefore, the main objectives of this study can be summarized as follows:

- To systematically review the existing literature on innovation and healthcare hotels;
- To explore the significance of medical hotels as innovative survival and growth strategies for the hospitality industry;
- To identify areas for future research in this field.

The chapter is structured as follows:

- Research methodology;
- Classification and analysis of the literature;

- Findings;
- Conclusion with future research prospects.

15.2 RESEARCH METHODOLOGY

The study adopted the systematic literature review (SLR) methodology to review and analyze articles. A search was conducted in the Google Scholar search engine for articles containing the terms "Hotel Innovation" or "Hospitality Innovation" and "Organizational Innovation" or "Medical Hotel." The initial 500 results, sorted by relevance according to Google Scholar, were considered. To ensure the quality of the studies, only articles published in peer-reviewed journals were included. The time horizon of the paper spans a decade (2009–2021), and the search was last conducted in December 2021. After applying the inclusion and exclusion criteria, a total of 55 articles were included for review.

15.3 CLASSIFICATION AND ANALYSIS OF THE LITERATURE

In this section, articles are classified based on publication year, journal title and citation analysis:

1. **Year of Publication:** Figure 15.1 shows the distribution of articles from 2009 to 2021. A positive trend in scholarly articles can be observed over the years.

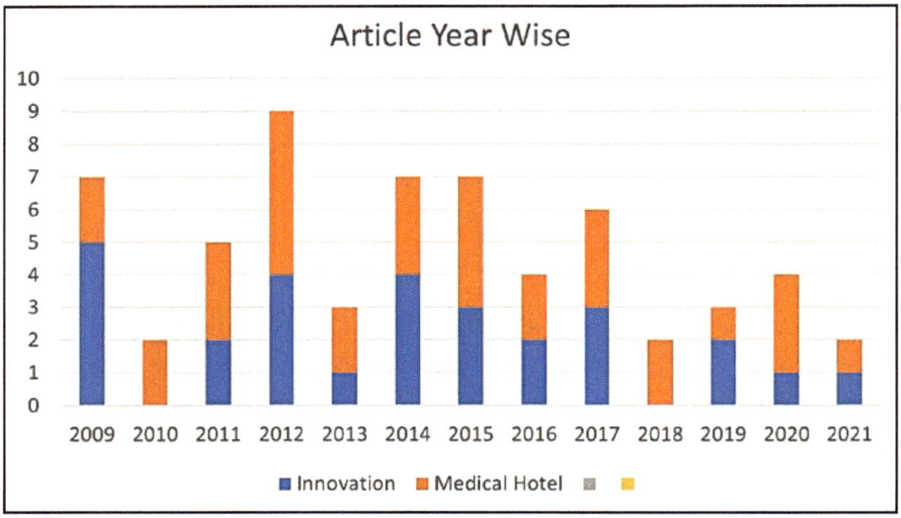

FIGURE 15.1 Article distribution 2009 to 2021.

2. **Journal of Publication:** Figure 15.2 shows the publication of the article in the journal. A total of 55 articles were taken from 32 journals. Only articles published in peer-reviewed journals were considered to ensure the quality of the literature. Out of those, eight journals have two or more publications, which are shown here, while the remaining journals have only one article published.

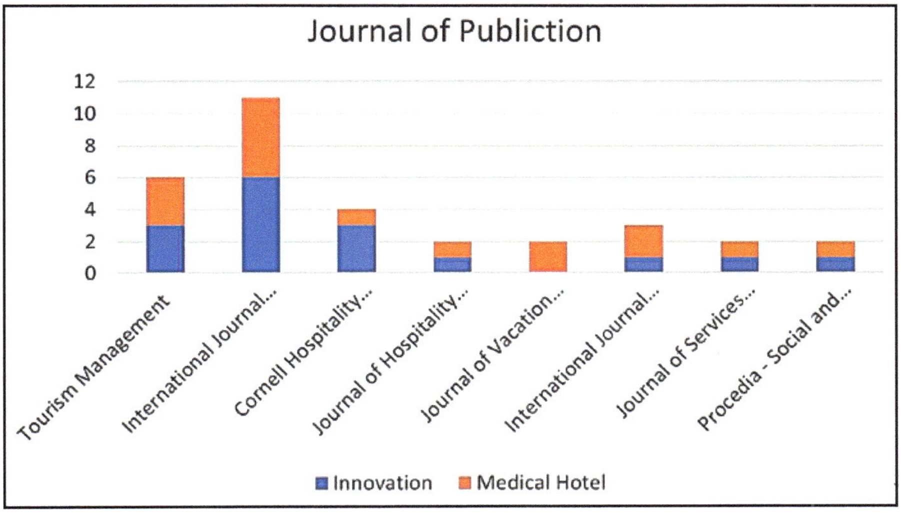

FIGURE 15.2 Article publication journal wise.

3. **Citation Analysis:** Here, the articles are analyzed based on their citation in other publications to determine their relevance in the study of innovation and medical hotels. The Google Scholar search engine is used for this purpose. Table 15.1 displays the articles along with their citation numbers. Only those articles with 80 or more citations are considered, in order to avoid a lengthy list.

TABLE 15.1 Article Listing According to Citations

Sl. No.	Article	Number of Citations
	Innovation	
1.	Han & Ryu (2009)	1,023
2.	Toivonen & Tuominen (2009)	638
3.	Orfila-Sintes & Mattsson (2009)	498

TABLE 15.1 *(Continued)*

Sl. No.	Article	Number of Citations
4.	Chen et al. (2009)	424
5.	Linea, & Runyan (2012)	268
6.	Martinez-Ros & Orfila-Sintes (2009)	250
7.	Becerra et al. (2013)	229
8.	Chen (2011)	215
9.	Tajeddini & Trueman (2012)	186
10.	Hilman & Kaliappen (2015)	96
11.	Concepción et al. (2011)	85
12.	Martin-Riosa & Ciobanu (2019)	81
13.	Sandvik et al. (2014)	80
	Medical Hotel	
14.	Heung et al. (2011)	351
15.	Grissemann et al. (2012)	279
16.	Gan & Fredrick (2011)	169
17.	Han (2013)	165
18.	Reddy et al. (2010)	157
19.	Konu et al. (2010)	137
20.	Han & Hwangb (2013)	99
21.	Bannister et al. (2009)	97
22.	López-Fernández et al. (2011)	85
23.	Heoa & Hyunb (2015)	84

15.4 FINDINGS AND DISCUSSION

Improvement of tangible factors and a customer-centric strategy have been identified as the most important factors in exceeding guest expectations (Ekinci et al., 2002; Victorino et al., 2005). It has also been found that a customer-driven perspective and openness to innovative ideas among employees are critical for the success of hospitality firms. However, it is observed that the hospitality industry is one of the least innovative service industries (Tajeddini & Trueman, 2012; Martin-Riosa & Ciobanu, 2019).

The concept of a medical hotel, which combines healthcare and the hospitality industry, has gained high acceptance among medical tourists and has been proven to have positive outcomes for patients. In addition to treatment, factors such as hygiene and safety, accommodation for family members, and proper food are found to be critical in choosing a medical hotel over a hospital or another hotel (Han, 2013; Shirzadi & Raeissi, 2016).

Medical hotels provide various non-medical services to patients, including tourist and medical packages. Proper marketing and branding of healthcare hotels are observed to be essential for their growth (Han & Hyun, 2014). The implementation of this innovative strategy, combining medical facilities with hotel amenities in the hospitality industry, aligns well with hotels already catering to customer needs (Hume & Demicco, 2007).

During pandemics or epidemics, when hotels struggle to survive, medical hotels have been found to be the best innovative strategy, as they can function as isolation centers (Ramírez-Cervantes et al., 2020; Breier et al., 2021). Finally, it has been observed that despite its potential, the concept of medical hotels remains relatively unknown in the hotel industry, resulting in a limited number of medical hotels worldwide (Han & Hyun, 2014).

15.5 CONCLUSION AND SCOPE FOR FUTURE RESEARCH

The objective of this chapter was to explore the concept of medical hotels as an innovative strategy through a SLR. Based on the literature, medical hotels can be concluded as an effective innovative strategy and the future of the hospitality, healthcare, and tourism industry. It is recommended that medical hotels be considered as a major marketing strategy in the hospitality industry (Jin & Zhang, 2016; Suess et al., 2020).

Despite the best efforts of the authors, there are limitations to this study that need to be addressed. Firstly, only articles written in English were considered, which may limit the scope of the findings. Secondly, the time frame used in the study, covering the period from 2000 to 2021, may not be sufficiently broad to capture all relevant research trends in hospitality and medical hotel innovation. Therefore, it is suggested to expand the time frame of the research to include a greater number of research articles and achieve more impactful outcomes.

The study brings the areas of future research which are as follows:

- First, future research can focus on the feasibility of medical hotels from the perspective of the hospitality industry, considering various criteria such as star category, property type, clientele type, and size.
- Second, further studies can be conducted by including a large sample of hotel employees.
- Lastly, additional research can be carried out on the professional collaboration between hotels and hospitals.

CONFLICTS OF INTEREST

The authors declare no conflicts of interest.

KEYWORDS

- **collaboration**
- **healthcare**
- **hospitality**
- **hotel industry**
- **innovation**
- **innovative strategy**
- **medical hotel**
- **tourism industry**

REFERENCES

Aladag, O., Köseoglu, M., King, B., & Mehraliyev, F., (2020). Strategy implementation research in hospitality and tourism: Current status and future potential. *International Journal of Hospitality Management, 88*, 102556. https://doi.org/10.1016/j.ijhm.2020.102556.

Backman, M., Klaesson, J., & O¨ ner, O., (2017). Innovation in the hospitality industry: Firm or location? *Tourism Economics, 23*(8), 1591–1614. doi: 10.1177/1354816617715159.

Bannister, B., Puro, V., Fusco, M., Hepstontell, J., & Ippolito, G., (2009). *Framework for the Design and Operation of High-Level Isolation Units: Consensus of the European Network of Infectious Disease, 9,* 45–56. www.thelancet.com/infection (accessed on 09 March 2024).

Becerra, M., Santaló, J., & Silva, R., (2013). Being better vs. being different: Differentiation, competition, and pricing strategies in the Spanish hotel industry. *Tourism Management, 34*, 71–79. http://dx.doi.org/10.1016/j.tourman.2012.03.014.

Bharwani, S., & Mathew, D., (2016). Customer service innovations in the Indian hospitality industry. *Worldwide Hospitality and Tourism Themes, 8*, 4. http://dx.doi.org/10.1108/Whatt-04-2016-0020.

Cervantes, K., Pardo, V., Tovar, C., Ales, G., & Diaz, Q., (2020). A medicalized hotel as a public health resource for the containment of COVID-19: More than a place of quarantining. *Journal of Public Health,* 1–9 https://doi.org/10.1093/pubmed/fdaa129.

Chen, J., & Tsou, H., (2009). Service delivery innovation antecedents and impact on firm performance. *Journal of Service Research, 12*(1), 36–55. 10.1177/1094670509338619.

Chen, J., (2011). Innovation in hotel services: Culture and personality. *International Journal of Hospitality Management, 30*, 64–72. https://doi: 10.1016/j.ijhm.2010.07.006.

D. Line, N., & Runyan, R., (2011). Hospitality marketing research: Recent trends and future directions. *International Journal of Hospitality Management, 31*(2), 477–488. https://doi.org/10.1016/j.ijhm.2011.07.006.

Docrates, (2011). *Healthcare Hotel Helsinki*. Retrieved from: http://www.docrates.com/en/docrates-hospital/health-care-hotel-helsinki (accessed on 09 March 2024).

Gan, L. L., & Frederick, J. R., (2011). Medical tourism facilitators: Pattern of service differentiation. *Journal of Vacation Marketing, 17*(3), 165–183.

GHN, (2011). *Global Healthcare Network*. Retrieved from: http://www.globalhealthcarenetwork.com/ (accessed on 09 March 2024).

Global Wellness Institute, (2017). *Global Wellness Economy Monitor*. https://globalwellnessinstitute.org/wpcontent/uploads/2018/06/GWI_WellnessEconomyMonitor2017_FINALweb.pdf (accessed on 09 March 2024).

Gomezel, D., (2016). A systematic review of research on innovation in hospitality and tourism. *International Journal of Contemporary Hospitality Management, 28*(3), 516–558. https://doi.org/10.1108/IJCHM-10-2014-0510.

Grisemann, U., Plank, A., & Brunner-Sperdin, A., (2012). Enhancing business performance of hotels: The role of innovation and customer orientation. *International Journal of Hospitality Management, 33*, 347–356. http://dx.doi.org/10.1016/j.ijhm.2012.10.005.

Guisado-González, M., Guisado-Tato, M., & Sandoval-Pérez, A., (2014). Determinants of innovation performance in Spanish hospitality companies: Analysis of the coexistence of innovation strategies. *The Service Industries Journal, 33*(6), 580–593. doi: 10.1080/02642069.2011.614343.

Han, H., & Hwang, J., (2013). Multi-dimensions of the perceived benefits in a medical hotel and their roles in international travelers' decision-making process. *International Journal of Hospitality Management, 35*, 100–108. https://doi.org/10.1016/j.ijhm.2013.05.011.

Han, H., & Hyun, S. S., (2014). Medical hotel in the growth of global medical tourism. *Journal of Travel and Tourism Marketing, 31*, 366–380. doi: 10.1080/10548408.2013.876955.

Han, H., & Ryu, K., (2009). The roles of the physical environment, price perception and customer satisfaction in determining customer loyalty in the restaurant industry. *Journal of Hospitality & Tourism Research, 33*, 487–510. https://doi.org/10.1177%2F1096348009344212.

Han, H., (2013). The healthcare hotel: Distinctive attributes for international medical travelers. *Tourism Management, 36*, 257–268. http://dx.doi.org/10.1016/j.tourman.2012.11.016.

Han, H., Kim, Y., Kim, C., & Ham, S., (2015). Medical hotels in the growing healthcare business industry: Impact of international traveler's perceived outcome. *Journal of Business Research, 68*(9), 1869–1877. https://doi.org/10.1016/j.jbusres.2015.01.015.

Haroon, S., Chandan, J. S., Midelton, J., & Cheng, K. K., (2020). COVID-19: Breaking the chain of household transmission. *BMJ, 370*, m3181. https://doi.org/10.1136/bmj.m3181.

Hoefer, A., Pampaka, D., Wagner, R., Herrera, A., Alonso, R., Perea, R., Porteco, C., et al., (2020). Management of a COVID-19 outbreak in a hotel in Tenerife, Spain. *International Journal of Infectious Disease, 96*, 384–386. https://doi.org/10.1016/j.ijid.2020.05.047.

Holte, H. H., Harboe, I., & Visit, G. E., (2011). *Effect of Patient Hotels for Hospitals and Patient*. Retrieved from: http://europepmc.org/article/NBK/NBK464817 (accessed on 09 March 2024).

Hung, K., Mark, C., Yeung, M., Chan, E., & Graham, A., (2018). The role of hotel industry in response to emerging epidemics: A case study of SARS in 2003 and H1N1 swine flu in 2009 in Hong Kong. *Globalization and Health, 14*, 117. https://doi.org/10.1186/s12992-018-0438-6.

Jin, L., & Zhang, R., (2016). Enlightenment of medical hotel on the development of medical tourism in China. *Advances in Social Science, Education and Humanities Research, 92*, 481–485. http://creativecommons.org/licenses/by-nc/4.0/ (accessed on 09 March 2024).

Kandampully, J., Keating, B., Kim, B., Mattila, A., & Solnet, D., (2014). Service research in the hospitality literature: Insights from a systematic review. *Cornell Hospitality Quarterly, 55*(3), 287–299. https://doi.org/10.1177/1938965514536778.

Kazehma, A., & Dehkordi, K., (2017). The principles of designing hospital hotel with the approach of health tourism in kish island. *Journal of History Culture and Art Research, 6*(1), 515–531. https://doi.org/10.7596/taksad.v6i1.758.

Kunwar, R. R., (2019). Medical tourism and hospitality in hospitals. *The Gaze Journal of Tourism and Hospitality, 10*, 67–123. https://doi.org/10.3126/gaze.v10i1.22778.

Langvinienơ, N., & Daunoraviþinjtơ, I., (2015). Factors influencing the success of business model in the hospitality service industry. *Procedia – Social and Behavioral Sciences, 213*, 902–910.

Larijani, B., & Zahedi, F., (2013). Patient centered care in dialectology: An Islamic perspective from Iran. *J. Diabet. Metabol. Discord, 12*, 18.

Lee, K. W., Choi, H. S., & Jeon, M. S., (2009). *Impact of the Medical Tourism Market on the Hotel Industry in South Korea: Focus on 5-Star Hotels*. Scholar works.

López-Fernández, M., Serrano-Bedia, A., & Gómez-López, R., (2011). Factors encouraging innovation in Spanish hospitality firms. *Cornell Hospitality Quarterly, 52*(2), 144–152. https://doi.org/10.1177/1938965510393723.

Lu, I. Y., & Shiu, J. Y., (2009). Customers' behavioral intentions in the service industry: An empirical study of Taiwan spa hotels. *The Asian Journal of Quality, 10*(3), 73–85. http://dx.doi.org/10.1108/15982680911021205.

Mackenzie, O. Z., Trimbur, M. C., & Vanjani, R., (2020). An isolation hotel for people experiences homelessness. *The New England Journal of Medicine*, 1–2. https://doi.org/10.1056/NEJMc2022860.

Martı´nez-Ros, E., & Orfila-Sintes, F., (2009). Innovation activity in the hotel industry. *Technovation, 29*, 632–641. doi: 10.1016/j.technovation.2009.02.004.

Martin-Rios, C., & Ciobanu, T., (2019). Hospitality innovation strategies: An analysis of success factors and challenges. *Tourism Management, 70*, 218–229. https://doi.org/10.1016/j.tourman.2018.08.018.

Medical Hotels, (2011). *Medical Hotel – The Health Care Hotels*. Retrieved from: http://www.medical-hotels.com/index.html (accessed on 09 March 2024).

Metaxas, I., Chatzoglou, P., & Koulouriotis, D., (2017). Proposing a new modus operandi for sustainable business excellence: The case of Greek hospitality industry. *Total Quality Management & Business Excellence*. doi: 10.1080/14783363.2017.1315934.

Nagy, A., (2014). The orientation towards innovation of spa hotel management: The case of Romanian spa industry. *Procedia – Social and Behavioral Sciences, 124*, 425–431.

Orfila-Sintes, F., & Mattsson, J., (2009). Innovation behavior in the hotel industry. *Omega, 37*, 380–394. https://doi.org/10.1016/j.omega.2007.04.002.

Pappas, N., (2015). Marketing hospitality industry in an era of crisis. *Tourism Planning & Development, 12*(3), 333–349. https://doi.org/10.1080/21568316.2014.979226.

Pena, M., Núnez-Serrano, J., Turrion, J., & Velazquez, F., (2016). Are innovations relevant for consumers in the hospitality industry? A hedonic approach for Cuban hotels. *Tourism Management, 55*, 184–196. https://doi.org/10.1016/j.tourman.2016.02.009.

Reddy, S. G., York, V. K., & Brannon, L. A., (2010). Travel for treatment: Student's perspective of medical tourism. *International Journal of Tourism Research, 12*, 510–522. https://doi.org/10.1002/jtr.769.

Salman, D., Tawfik, Y., Samy, M., & Artal-Tur, A., (2017). A new marketing mix model to rescue the hospitality industry: Evidence from Egypt after the Arab spring. *Future Business Journal, 3*(1), 47–69. https://doi.org/10.1016/j.fbj.2017.01.004.

Sandvik, I., Duhan, D., & Sandvik, K., (2014). Innovativeness and profitability: An empirical investigation in the Norwegian hotel industry. *Cornell Hospitality Quarterly, 55*(2) 165–185. doi: 10.1177/1938965514520963.

Shirzadi, S. M., Raeissi, P., Nasiripour, A. A., & Tabibi, S. J., (2016). Factors affecting the quality of hospital hotel services from the patients and their companions' point of view: A national study in Iran. *Journal of Research in Medical Sciences, 21*, 46. https://doi.org/10.4103/1735-1995.183987.

Sive, J., Ardeshna, M., Chessman, S., Grange, L., Morris, S., Nicholas, C., Peggs, K., et al., (2012). Hotel based ambulatory care for complex cancer patients: A review of the University College London Hospital experience. *Leukemia & Lymphoma, 53*(12), 2397–2404. https://doi.org/10.3109/10428194.2012.694430.

Steele, J. R., Jones, A., Clarke, R., & Shoemaker, S., (2015). Healthcare delivery meets hospitality: A pilot study in radiology. *Journal of The American College of Radiology, 12*(6), 587–593. https://doi.org/10.1016/j.jacr.2014.10.008.

Suess, C., & Mody, M., (2017). Hospitality healthscape: A conjoint analysis approach to understanding patient responses to hotel-like hospital rooms. *International Journal of Hospitality Management, 61*, 59–72. http://dx.doi.org/10.1016/j.ijhm.2016.11.004.

Tajeddini, K., & Trueman, M., (2012). Managing Swiss hospitality: How cultural antecedents of innovation and customer-oriented value systems can influence performance in the hotel industry. *International Journal of Hospitality Management, 31*, 1119–1129. doi: 10.1016/j.ijhm.2012.01.009.

Zygourakis, C., Rolston, D., Treadway, J., Chang, S., & Kilot, M., (2014). What do hotels and hospitals have in common? How can we learn from the hotel industry to take better care of patients? *Surgical Neurology International, 4*(5), S49–S53. doi: 10.4103/2152-7806.128913.

Index

J

K

L

M

N

T